P9-DNN-507

Encyclopedia of EARTH AND PHYSICAL SCIENCES

Second Edition

8

Pine Ridge School Library
9505 Williston Road
Williston, VT 05495

Nanotechnology – Paleoclimatology

New York

EDITORIAL BOARD

Consultants-in-chief
Frank G. Ethridge, Professor of Geology, Colorado State University, Boulder, Colorado
Donald R. Franceschetti, Distinguished Service Professor, University of Memphis, Memphis, Tennessee

Chemistry
Professor Dennis Barnum, Portland State University, Portland, Oregon
Professor Peter Bridson, University of Memphis, Memphis, Tennessee
Martin Clowes, M.A. (University of Oxford, Oxford, England)
Dr. Mark Freilich, University of Memphis, Memphis, Tennessee
Adrian Hall, Heriot-Watt University, Edinburgh, Scotland
Andrew Mercer, Ph.D. (University of the Witwatersrand, Johannesburg, South Africa)
Professor William Motherwell, University College London, London, England
Christine Rae, University of Edinburgh, Edinburgh, Scotland
Craig von Dolleweerd, Guy's Hospital, London, England

Geology and mineralogy
Dr. Cassandra Coombs, College of Charleston, Charleston, South Carolina
Marc Coyne, University of Kentucky, Lexington, Kentucky
Douglas Palmer, Ph.D. (Trinity College, Dublin, Ireland)
Chris Pellant, B.A. (Keele University, Keele, England)
Helen Pellant, B.A. (Leeds University, Leeds, England)
Dr. Michael Romano, University of Sheffield, Sheffield, England
Dr. Norman Savage, University of Oregon, Eugene, Oregon

Mathematics
John Bolgar, D.Phil. (University of Oxford, Oxford, England)
Dr. Martin Hazleton, University College London, London, England
Professor Jonathan Leech, Westmont College, Santa Barbara, California
Nathan Lepora, Ph.D. (University of Cambridge, Cambridge, England)
Dr. Leonard Smith, University of Oxford, Oxford, England

Meteorology
Jane Insley, Science Museum, London, England
Dr. James Norwine, Texas A & M University, Kingsville, Texas

Oceanography
Sharon Roth Franks, Ph.D., University of California, San Diego, California
Barbara Ransom, National Science Foundation, Arlington, Virginia
Richard Strickland, University of Washington, Seattle, Washington

Physics and space science
Peter Bond, B.A. (University of London, London, England)
Professor George Flynn, State University of New York, Plattsburgh, New York
Marc D. Rayman, Ph.D., NASA/JPL, California Institute of Technology, Pasadena, California
Professor Ernst von Meerwall, University of Akron, Akron, Ohio

Marshall Cavendish
99 White Plains Road
Tarrytown, New York 10591-9001

www.marshallcavendish.us
© 1998, 2006 Marshall Cavendish Corporation

All rights reserved. No part of this book may be reproduced or utilized in any form or by any means electronic or mechanical, including photocopying, recording, or by any information storage and retrieval system, without prior written permission from the copyright holders.

Library of Congress Cataloging-in-Publication Data

Encyclopedia of earth and physical sciences.— 2nd ed.
p. cm.
Includes bibliographical references and indexes.
Contents: 1. Absolute zero–Barrier islands—2. Base level–Clouds—3. Cloud seeding–Earth, structure of—4. Earthquakes–Forests—5. Fossil record–Humidity—6. Hurricanes–Magnetism—7. Maps and mapping–Musical acoustics—8. Nanotechnology–Paleoclimatology—9. Paleontology–Quasars—10. Quaternary period–Space—11. Space exploration–Tin—12. Titanium–Zinc—13. Index volume.
ISBN 0-7614-7583-4 (set : alk. paper) — ISBN 0-7614-7584-2 (v. 1 : alk.paper)—ISBN 0-7614-7585-0 (v. 2 : alk. paper)—ISBN 0-7614-7586-9 (v. 3 : alk. paper)—ISBN 0-7614-7587-7 (v. 4 : alk. paper)—ISBN 0-7614-7588-5 (v. 5 : alk. paper)—ISBN 0-7614-7589-3 (v. 6 : alk.paper)—ISBN 0-7614-7590-7 (v. 7 : alk. paper)—ISBN 0-7614-7591-5 (v. 8 : alk. paper)—ISBN 0-7614-7592-3 (v. 9 : alk. paper)—ISBN 0-7614-7593-1 (v. 10 : alk. paper)—ISBN 0-7614-7594-X (v. 11 : alk. paper)—ISBN 0-7614-7595-8 (v. 12 : alk. paper)—ISBN 0-7614-7596-6 (v. 13 : alk. paper)
1. Earth sciences—Encyclopedias. 2. Physical sciences—Encyclopedias. I. Marshall Cavendish Corporation.

QE5.E513 2005
500.2'03--dc22

2004058630

Printed in Malaysia

09 08 07 06 05 1 2 3 4 5

Cover illustration: view of Earth and the Moon from Galileo spacecraft (NASA)
Title page illustration: rusty iron pipes (Photos.com)

Marshall Cavendish
Editor: Joyce Tavolacci
Editorial Director: Paul Bernabeo
Production Manager: Michael Esposito

Created by **The Brown Reference Group plc**
Project Editor: Lesley Campbell-Wright
Editor: Tom Jackson
Designers: Joan Curtis, Stefan Morris, and Sarah Williams
Picture Researcher: Susy Forbes
Managing Editor: Tim Harris
Design Manager: Lynne Ross
Indexer: Kay Ollerenshaw

PHOTOGRAPHIC CREDITS

American Institute of Physics: Choi *1014,* Hongjie Dai *1015,* Dustin W. Carr and Harold G. Craighead *1013;* **Antares:** Francois Montanet *1037;* **ArgusLab 4.0:** Mark A. Thompson *1128;* **Corbis:** *1046,* Dave Bartruff *1029,* Jonathan Blair *1104,* Adam Woolfitt *1132;* **Frank Lane Picture Agency:** *1022, 1048, 1119, 1143;* **Frank Spooner:** *1093;* **Geoscience Features:** *1115;* **Getty Images:** *1083;* **Marshall Cavendish:** *1072;* **Mary Evans Picture Library:** *1039, 1054, 1085, 1088, 1091, 1098, 1121;* **NASA:** *1040, 1042, 1043, 1147;* **Photodisc:** *1082;* **Photos.com:** *1064;* **Science and Society:** *1125,* Brian and Mavis Bousfield *1100, 1101,* NMPFT Daily Herald Archive *1041,* Science Museum *1028, 1060;* **SPL:** *1016–1018, 1020, 1021, 1024, 1031–1034, 1036, 1047, 1051, 1053, 1055–1059, 1061–1063, 1067, 1069, 1070, 1071, 1073–1075, 1077–1081, 1086, 1089, 1094, 1096, 1097, 1099, 1105, 1107–1109, 1111–1114, 1116–1118, 1120, 1122, 1123, 1134, 1135, 1137, 1139–1142, 1144–1146;* **Topham:** *1045,* Larry Mulvehill/The Image Works *1023,* Frank Pedrick/The Image Works *1103;* **TRH:** *1133,* Crumman *1019,* NASA *1035;* **University of Pennsylvania:** Smith Collection *1044;* **University of Sheffield, UK:** Dr. R. Edyvean *1102;* **US Defense:** *1026, 1030;* **US Mint:** *1052.*

ARTWORK CREDITS
Darren Awuah: *1065, 1130, 1131, 1148.*
Bill Botten: *1025, 1038, 1049, 1131, 1138.*
Jennie Dooge: *1048, 1076, 1090, 1095, 1106, 1110, 1136.*

CONTENTS

PERIODIC TABLE

Key: 1 / **H** / 1.00794 — Atomic number (top), Atomic mass (bottom)

Metals | Metalloids | Nonmetals

Periods	1 (IA)	2 (IIA)	3 (IIIB)	4 (IVB)	5 (VB)	6 (VIB)	7 (VIIB)	8	9 (VIII)	10	11 (IB)	12 (IIB)	13 (IIIA)	14 (IVA)	15 (VA)	16 (VIA)	17 (VIIA)	18 (0)
1	1 **H** 1.00794																	2 **He** 4.00260
2	3 **Li** 6.941	4 **Be** 9.01218											5 **B** 10.811	6 **C** 12.011	7 **N** 14.0067	8 **O** 15.9994	9 **F** 18.9984	10 **Ne** 20.1797
3	11 **Na** 22.9897	12 **Mg** 24.3050											13 **Al** 26.98154	14 **Si** 28.0855	15 **P** 30.97376	16 **S** 32.066	17 **Cl** 35.4527	18 **Ar** 39.948
4	19 **K** 39.0983	20 **Ca** 40.078	21 **Sc** 44.95591	22 **Ti** 47.88	23 **V** 50.9415	24 **Cr** 51.9961	25 **Mn** 54.9380	26 **Fe** 55.847	27 **Co** 58.93320	28 **Ni** 58.6934	29 **Cu** 63.546	30 **Zn** 65.39	31 **Ga** 69.723	32 **Ge** 72.61	33 **As** 74.9215	34 **Se** 78.96	35 **Br** 79.904	36 **Kr** 83.80
5	37 **Rb** 85.4678	38 **Sr** 87.62	39 **Y** 88.90585	40 **Zr** 91.224	41 **Nb** 92.90638	42 **Mo** 95.94	43 **Tc** 98.9072	44 **Ru** 101.07	45 **Rh** 102.9055	46 **Pd** 106.42	47 **Ag** 107.8682	48 **Cd** 112.411	49 **In** 114.82	50 **Sn** 118.710	51 **Sb** 121.76	52 **Te** 127.60	53 **I** 126.9045	54 **Xe** 131.29
6	55 **Cs** 139.9054	56 **Ba** 137.327	57 ***La** 138.9055	72 **Hf** 178.49	73 **Ta** 180.9479	74 **W** 183.85	75 **Re** 186.207	76 **Os** 190.2	77 **Ir** 192.22	78 **Pt** 195.08	79 **Au** 196.9665	80 **Hg** 200.59	81 **Tl** 204.3833	82 **Pb** 207.2	83 **Bi** 208.98045	84 **Po** 208.9824	85 **At** 209.9871	86 **Rn** 222.0176
7	87 **Fr** 223.0197	88 **Ra** 226.0254	89 **†Ac** 227.0278	104 **Rf** (263)	105 **Db** (268)	106 **Sg** (266)	107 **Bh** (272)	108 **Hs** (277)	109 **Mt** (276)	110 **Ds** (281)	111 **Rg** (280)	112 **Uub** (285)	113 **Uut** (284)	114 **Uuq** (289)	115 **Uup** (288)	116 **Uuh** (292)		

*	58 **Ce** 140.115	59 **Pr** 140.9076	60 **Nd** 144.24	61 **Pm** 144.9127	62 **Sm** 150.36	63 **Eu** 151.965	64 **Gd** 157.25	65 **Tb** 158.9254	66 **Dy** 162.50	67 **Ho** 164.9303	68 **Er** 167.26	69 **Tm** 168.9342	70 **Yb** 173.04	71 **Lu** 174.967
†	90 **Th** 232.0381	91 **Pa** 231.0359	92 **U** 238.0289	93 **Np** 237.0482	94 **Pu** 244.0642	95 **Am** 243.0614	96 **Cm** 247.0703	97 **Bk** 247.0703	98 **Cf** 251.0796	99 **Es** 252.083	100 **Fm** 257.0951	101 **Md** 258.10	102 **No** 259.1009	103 **Lr** 262.11

Actinium (89)	Ac
Aluminum (13)	Al
Americium (95)	Am
Antimony (51)	Sb
Argon (18)	Ar
Arsenic (33)	As
Astatine (85)	At
Barium (56)	Ba
Berkelium (97)	Bk
Beryllium (4)	Be
Bismuth (83)	Bi
Bohrium (107)	Bh
Boron (5)	B
Bromine (35)	Br
Cadmium (48)	Cd
Calcium (20)	Ca
Californium (98)	Cf
Carbon (6)	C
Cerium (58)	Ce
Cesium (55)	Cs
Chlorine (17)	Cl
Chromium (24)	Cr
Cobalt (27)	Co
Copper (29)	Cu
Curium (96)	Cm
Darmstadtium (110)	Ds
Dubnium (105)	Db
Dysprosium (66)	Dy
Einsteinium (99)	Es
Erbium (68)	Er
Europium (63)	Eu
Fermium (100)	Fm
Fluorine (9)	F
Francium (87)	Fr
Gadolinium (63)	Gd
Gallium (31)	Ga
Germanium (32)	Ge
Gold (79)	Au
Hafnium (72)	Hf
Hassium (108)	Hs
Helium (2)	He
Holmium (67)	Ho
Hydrogen (1)	H
Indium (49)	In
Iodine (53)	I
Iridium (77)	Ir
Iron (26)	Fe
Krypton (36)	Kr
Lanthanum (57)	La
Lawrencium (103)	Lr
Lead (82)	Pb
Lithium (3)	Li
Lutetium (71)	Lu
Magnesium (12)	Mg
Manganese (25)	Mn
Meitnerium (109)	Mt
Mendelevium (101)	Md
Mercury (80)	Hg
Molybdenum (42)	Mo
Neodymium (60)	Nd
Neon (10)	Ne
Neptunium (93)	Np
Nickel (28)	Ni
Niobium (41)	Nb
Nitrogen (7)	N
Nobelium (102)	No
Osmium (76)	Os
Oxygen (8)	O
Palladium (46)	Pd
Phosphorus (15)	P
Platinum (78)	Pt
Plutonium (94)	Pu
Polonium (84)	Po
Potassium (19)	K
Praseodymium (59)	Pr
Promethium(61)	Pm
Protactinium (91)	Pa
Radium (88)	Ra
Radon (86)	Rn
Rhenium (75)	Re
Rhodium 45)	Rh
Roentgenium (111)	Rg
Rubidium (37)	Rb
Ruthenium (44)	Ru
Rutherfordium (104)	Rf
Samarium 62)	Sm
Scandium (21)	Sc
Seaborgium (106)	Sg
Selenium (34)	Se
Silicon (14)	Si
Silver (47)	Ag
Sodium (11)	Na
Strontium (38)	Sr
Sulfur (16)	S
Tantalum (73)	Ta
Technetium (43)	Tc
Tellurium (52)	Te
Terbium (65)	Tb
Thallium (81)	Tl
Thorium (90)	Th
Thulium (69)	Tm
Tin (50)	Sn
Titanium (22)	Ti
Tungsten (74)	W
Ununbium (112)	Uub
Ununhexium (116)	Uuh
Ununtrium (113)	Uut
Ununpentium (115)	Uup
Ununquadrium (114)	Uuq
Uranium (92)	U
Vanadium (23)	V
Xenon (54)	Xe
Ytterbium (70)	Yb
Yttrium (39)	Y
Zinc (30)	Zn
Zirconium (40)	Zr

USEFUL INFORMATION

PHYSICAL CONSTANTS AND SOLAR SYSTEM DATA

Quantity	Symbol	Value	SI Unit
Avogadro's number	N_A	6.023×10^{23}	mol^{-1}
Boltzmann's constant	k	1.381×10^{-23}	JK^{-1}
Elementary charge	e	1.602×10^{-19}	C
Gravitational constant	G	6.673×10^{-11}	$m^3kg^{-1}s^{-2}$
Lunar mass	M_m	7.35×10^{22}	kg
Lunar radius (mean)	R_m	1.74×10^{6}	m
Mass of Earth	—	5.98×10^{24}	kg
Rest mass of electron	m_e	9.109×10^{-31}	kg
Rest mass of neutron	m_n	1.675×10^{-27}	kg
Rest mass of proton	m_p	1.673×10^{-27}	kg
Molar gas constant	R	8.314	$J(mol\,K)^{-1}$
Permeability of vacuum	μ_o	$4\pi \times 10^{-7}$	TmA^{-1}
Permittivity of vacuum	ε_0	8.854×10^{-12}	Fm^{-1}
Planck's constant	h	6.626×10^{-34}	Js
Radius of Earth (at equator)	—	6.34×10^{6}	m
Solar mass	$M_{\odot}$	1.989×10^{30}	kg
Solar radius (mean)	$R_{\odot}$	6.960×10^{9}	m
Speed of light in vacuum	c	2.998×10^{8}	ms^{-1}
Standard gravity	g	9.80665	ms^{-2}

SI BASE UNITS

Quantity	Unit	Symbol
Length	meter	m
Mass	kilogram	kg
Time	second	s
Electric current	ampere	A
Thermodynamic temperature	kelvin	K
Amount of substance	mole	mol
Luminous intensity	candela	cd

SI PREFIXES

Prefix	Symbol	Multiplier
Exa-	E	10^{18}
Peta-	P	10^{15}
Tera-	T	10^{12}
Giga-	G	10^{9}
Mega-	M	10^{6}
Kilo-	k	10^{3}
Hecto-	h	10^{2}
Deca-	da	10^{1}
Deci-	d	10^{-1}
Centi-	c	10^{-2}
Milli-	m	10^{-3}
Micro-	μ	10^{-6}
Nano-	n	10^{-9}
Pico-	p	10^{-12}
Femto-	f	10^{-15}
Atto-	a	10^{-18}

SI DERIVED UNITS

Quantity	Unit	Symbol	Equivalent Units
Absorbed dose	gray	Gy	Jkg^{-1}
Activity (of radionuclides)	becquerel	Bq	s^{-1}
Capacitance	farad	F	CV^{-1}
Electric charge	coulomb	C	As
Electric potential	volt	V	JC^{-1}
Electric resistance	ohm	Ω	VA^{-1}
Energy	joule	J	Nm
Force	newton	N	$kgms^{-2}$
Frequency	hertz	Hz	s^{-1}
Magnetic flux density	tesla	T	$kgA^{-1}s^{-2}$
Power	watt	W	Js^{-1}
Pressure	pascal	Pa	Nm^{-2}

Note: SI refers to *Système international d'unités*, or International System of Units, commonly called the metric system.

NANOTECHNOLOGY

A method of precisely positioning atoms and molecules to produce entirely new materials and machines

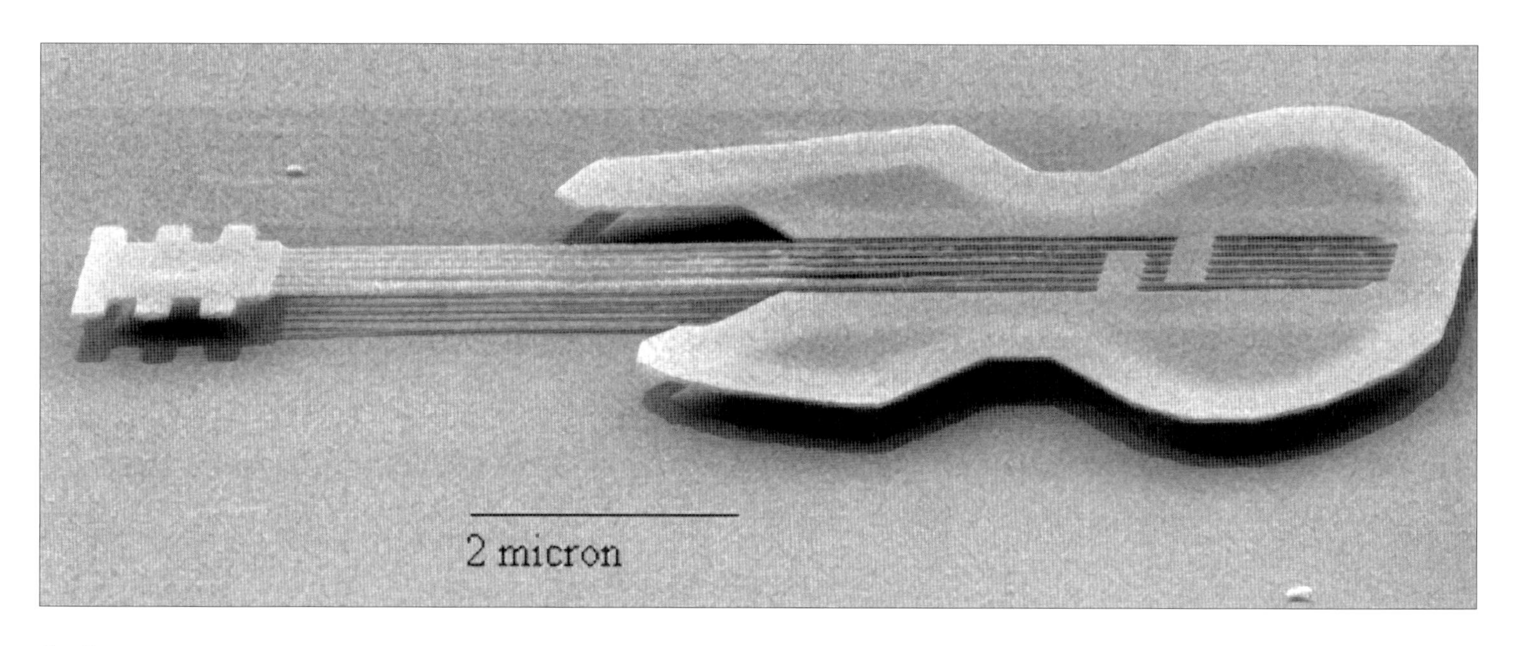

The world's smallest guitar, pictured above, is 10 micrometers long, smaller than a red blood cell and twenty times smaller than the width of a human hair. The sound produced by plucking the guitar's strings is too high-pitched to be audible.

Much of modern-day physical science is based on a simple assumption: atoms and molecules are the basic building blocks of matter (see ATOMS). The same atoms, when rearranged, can make very different materials: carbon atoms arranged one way make graphite (pencil lead); assembled in a different way, they make diamond (see CARBON). It follows that if scientists could assemble atoms and molecules to their own designs, they could make virtually anything—brand new medicines, computer chips, and all kinds of other things that have never previously been possible. They could also use nanomachines (tiny machines made as described above), sometimes called fabricators or assemblers, to build or repair ordinary-size objects. Using atoms and molecules like tiny building blocks, is known as nanotechnology or molecular manufacturing.

Nanotechnology takes its name from *nano*, a prefix that means "billionth." A nanometer (nm; one-billionth of a meter) is equal to 0.000000001 m (see MEASUREMENT). If a microscope could magnify a nanometer so it became the same length as a pencil, a pencil viewed at the same scale would be 150 miles (240 km) long. In nanotechnology things generally happen on a scale the size of molecules, which range from about one-tenth of a nanometer to thousands of nanometers in length—the so-called nanoscale. This scale is about a thousand times smaller than microscopic, where things happen on a scale of micrometers (μm; millionths of a meter).

Working on the nanoscale presents an obvious difficulty: atoms and molecules are far too small for people to see or manipulate using conventional tools. Instead, nanotechnologists rely on scanning probe microscopes (SPMs), which, as their name suggests, work by scanning a very sharp, very small probe across the surface of a sample. This device either generates images of the molecular world inside it or moves its atoms and molecules around. SPMs can nudge molecules, roll them along, or even pick them up from one place and put them down in another.

Some scientists dismiss nanotechnology as a fantasy, but arguably, life itself is proof that nanotechnology is viable. The chemistry and biology of the natural world have revealed many examples of "machines" built from molecules. One of the best-known examples is the so-called flagellum motor produced by bacteria such as *E.coli*. When these bacteria need to eat, they use proteins to build a spiral tail called a flagellum and whip it around at speed to move them toward a food source.

According to nanotechnologists, if nature can produce such elegant nanomachines, so can humans. However, scientists will have to learn the lessons of nature to produce nanomachines of any size. Because nanotechnology is so small and hard to produce, nanomachines will need to be able to reproduce themselves (replicate).

History

The basic idea of nanotechnology was first outlined by Nobel prize winning U.S. physicist Richard Feynman (1918–1988) at an after-dinner speech in 1959. His vision became a practical possibility in the

CORE FACTS

- Nanotechnologists work on a scale billionths of a meter using high-powered microscopes.
- Nanomaterials are already in common use, but nanomachines are still in the very early stages of development.
- Many scientists believe social and environmental impacts must be carefully evaluated before nanotechnology is adopted on a large scale.

CONNECTIONS

- Nanotechnology could revolutionize **MATERIALS SCIENCE** through the development of new composites, plastics, and other materials.
- Critics fear runaway nanotechnology could destroy the **BIOSPHERE**.

CARBON NANOTUBES

Nanotubes astonished scientists when they were discovered in 1991. They are carbon molecules that exist in the shape of a hollow tube, approximately one nanometer in diameter, whose wall can be as thin as a single atom (see FULLERENES). Despite their size, they are incredibly strong. Able to be grown to any length, effectively, they are the ultimate carbon fiber. Although their structure is quite different from any other molecule, they do have something in common with graphite, one of the other molecular forms (allotropes) of carbon. Like graphite, nanotubes can be used to make very durable nanoscopic "pencils." In 1998 a group of U.S. scientists used a nanotube to write the words "nanotube nanopencil" onto a piece of silicon with marks only 10 nanometers wide. It was the smallest piece of writing in human history.

A CLOSER LOOK

early 1980s when IBM scientists working in Zurich, Switzerland, developed the first SPM. In 1991 Japanese scientist Sumio Iijima (b. 1939) discovered a new form of carbon, the nanotube, which was soon being used as one of the basic building blocks to make nanomaterials and nanomachines (see FULLERENES). Much of the current media interest in nanotechnology stems from the work of Eric Drexler (b. 1955), whose 1986 book *Engines of Creation* considered how nanotechnology could lead to huge changes in virtually every aspect of society. Public interest prompted by such books has led to huge government investments in nanotechnology. The U.S. National Nanotechnology Initiative (NNI), launched by President Bill Clinton in 2000, is investing billions of dollars in nanotechnology and estimates that it will become a $1 trillion industry by 2020.

This small color screen creates a display using carbon nanotubes. Like all fullerenes, nanotubes are aromatic molecules (that is, they have delocalized electrons). This trait allows them to conduct electricity and be used in equipment such as the device below.

Applications

Although a great deal of nanotechnology remains highly speculative, revolutionary nanomaterials, including magnetic recording tapes, flexible plastics, and computer chips, are already available. Some clothing companies sell stain-resistant garments coated with "nanowhiskers," nanoscopic surface fibers that prevent stains from penetrating the fabric. One of the best-selling nano-products is sunscreen, which uses nanoparticles of zinc oxide to filter out harmful ultraviolet radiation from sunlight to stop it from reaching the skin. The Toyota company has developed nanocomposites (materials that combine two other materials) from plastics and clay to make a range of automobile parts, including lightweight, scratch-resistant bumpers.

In the microelectronic industry, nanotechnology is helping computer designers to increase the speed and power of their machines by making it possible to pack much greater numbers of electronic components onto much small microchips. In 1998 scientists made a transistor (a miniature electronic switch) out of a single carbon nanotube. That invention made it possible to use carbon nanotubes as the basic components of flatscreen computer and TV displays.

Much of the popular interest in nanotechnology stems not from nanomaterials such as composites and microchips but from nanomachines assembled from traditional machine parts, such as cogs, wheels, and levers, that have been built out of nanotubes. The most advanced nanostructures produced so far include a basic gear, a simple pump, a motion controller, and a simple switch. These components might one day be used to make tiny motors to power nanoscale robots, which could be sent into larger machines where people cannot go (for example, inside a nuclear power plant) or even into cells of the human body to carry out repairs.

Issues

Despite great progress in nanotechnology, "nanobots" and nanomachines might not become a reality for decades. One reason is the sheer difficulty of manipulating small structures, even under SPM microscopes. Different rules of physics come into play at the nanoscale. One major problem is that when atoms and molecules have been positioned with great precision, they can be suddenly thrown apart by thermal noise (vibrations caused by heat energy). In 1989 an IBM researcher named Don Eigler used a high-powered microscope to rearrange the positions of 35 xenon atoms so they spelled out the letters *IBM*. It was a spectacular demonstration of nanotechnology, but it took Eigler 22 hours to complete, and he had to work at −454°F (−270°C), a temperature approaching absolute zero (theoretically the lowest possible temperature that can exist; see HEAT); had it been any hotter, thermal vibration

These words have been written on a wafer of silicon dioxide that is only 10 nm wide. The words were written with a "nanopencil." This device is made up of a silicon lance with a carbon nanotube tip.

would have jiggled the xenon atoms out of place (see BROWNIAN MOTION).

Practical difficulties are only one of the stumbling blocks that nanotechnologists face. Much of the media coverage of the subject has focused on apocalyptic scare stories, including the risks posed by out-of-control nanobots (see box below). There are concerns that nanomaterials could pose environmental dangers and that they should be subject to the same strict assessments as genetically modified organisms and other new forms of biotechnology. Microscopic carbon particles, produced by vehicles and during coal mining, are known to play a role in serious respiratory disorders. Could the same harm be possible with even smaller carbon nanotubes?

The champions of nanotechnology herald this new field as nothing less than the next industrial revolution, suggesting it could solve virtually all of humankind's most pressing problems, from ending world poverty and curing cancer to developing superfast, supersmall computers and affordable space travel. Some scientists have even proposed using nanotubes to build a gigantic "elevator" that could carry humans and cargo from Earth into space. Whether or not these predictions are credible, it is likely to be many years before even the most modest of them comes true. Until then, nanotechnology will remain one of the most controversial technological developments of the 21st century.

C. WOODFORD

See also: ATOMS; BROWNIAN MOTION; CARBON; FULLERENES; HEAT; MEASUREMENT

Further reading

The Editors of *Scientific American*. 2002. *Understanding Nanotechnology*. New York: Warner Books.

Ratner, Mark, and Daniel Ratner. 2003. *Nanotechnology: A Gentle Introduction to the Next Big Idea*. Upper Saddle River, N.J.: Prentice-Hall.

U.S. National Nanotechnology Initiative: http://www.nano.gov

GRAY GOO

Nanotechnologists dream of creating nanobots, nanoscale robots that could revolutionize industry and medicine. Yet one of the most basic assumptions the researchers have made is that nanomachines should be self-replicating (able to reproduce themselves). So what happens if nanobots get out of control, reproduce themselves malignantly like cancers or viruses, and start destroying the humans who created them or the environment on which people depend? Eric Drexler, a leading author and researcher in the area, famously called this the "gray goo" problem, because that's all that would be left of the biosphere if nanobots gobbled it up. Drexler's alarming vision led many people to argue that nanotechnology must be developed with more caution. Some have gone so far as to argue that such controversial research should be suspended. However, others maintain that the fears are overstated.

SCIENCE AND SOCIETY

NANOMEDICINE

Modern drug treatments work by introducing chemicals into the body to correct basic biological processes that have, for some reason, gone awry. However, in the future, nanotechnology could cure illnesses and diseases in a quite different way. Nanobots released from a small, easy-to-swallow tablet could patrol the human body like a submicroscopic medical army, seeking out and destroying viruses, repairing DNA, cleaning out arteries, and killing cancer cells. Some researchers believe nanobots could even be used to reverse the aging process so that humans could live forever. If these forecasts come true, nanomedicine could prove to be one of the most lucrative areas of nanotechnology, but most of its benefits are likely to be many decades in the future.

LOOKING TO THE FUTURE

NASA

NASA is a U.S. government agency that works on aeronautical and space research and technological activities

The National Aeronautics and Space Administration (NASA) is a U.S. government agency that oversees aeronautical and space research and technology development activities that are not principally concerned with defense.

The National Advisory Committee for Aeronautics (NACA) was set up in 1915. Although it began as a small agency to "supervise and direct the study of the problems of flight," by the end of World War II (1939–1945), it had become the leading aeronautics research organization in the United States.

In 1957 the Soviet Union successfully launched *Sputnik 1*, the first artificial satellite, and the U.S. government realized that it was essential to set up a stronger space research program. In 1958 NASA was established to replace NACA. Whereas NACA had concentrated on collecting and interpreting data for others to use in aeronautical development, NASA was given a much broader agenda.

Over the next decades, the Soviet Union and the United States, participating in a so-called space race, launched thousands of satellites and probes.

Mission Control at NASA's Jet Propulsion Laboratory, Pasadena, California. The computer display in the background shows a close-up of the giant red spot in Jupiter's atmosphere, taken during one of the Voyager orbits.

The main objectives of NASA

NASA's principal goal—defined as seeking and encouraging, to the maximum extent possible, the "commercial use of space"—was to be carried with the following objectives:

- Expand human knowledge of Earth, the atmosphere, and space.
- Improve "the usefulness, performance, speed, safety, and efficiency of aeronautical and space vehicles."
- Develop and operate vehicles to carry equipment, plants, and animals through space.
- Establish studies of potential benefits and problems, together with opportunities for using aeronautics and space for peaceful and scientific purposes.
- Preserve "the role of the United States as a leader in aeronautical and space science and technology."
- Share all discoveries with U.S. defense agencies.
- Cooperate with "other nations and groups of nations in peaceful ventures in aeronautics and space."
- Use the best science and engineering resources in the United States, and cooperate with all agencies to avoid duplication of effort, facilities, and equipment.

These objectives have led NASA to also conduct research on ground propulsion systems, energy resource development, and bioengineering.

NASA and astronautics

From the beginning, NASA has always had to juggle government budget limitations, deadlines, political goals, and public interest with the roles of research. Since experiments are designed to test new ideas or gather information that may eventually expand knowledge or lead to an invention, setbacks or unexpected results inevitably have to be taken into account. The space race spotlighted NASA's successes in astronautic research and made any failures a cause of public concern. NASA's nonastronautic research was given far less attention.

NASA focused much of its efforts on developing and improving technology to propel people and spacecraft into space and to allow them to operate there. There have been many major programs associated with NASA's astronautics, including the Mercury program, the Gemini program, the Apollo program, the shuttle program, and the International Space Station.

CORE FACTS

- NASA was established in 1958, initially to compete with the former Soviet Union in the space race.
- The agency's goals have been defined on a broad basis and are designed to improve human life on Earth.
- NASA has designed and built spacecraft that have carried astronauts to the Moon and made uncrewed flights to all the planets except Pluto, many of the moons of the planets, the Sun, and several asteroids and comets.
- NASA led the way in developing satellites that now provide a wealth of information for mapmakers, oil and mineral prospectors, ecologists, and geographers.

CONNECTIONS

- Neil Armstrong was the first person on the **MOON**.
- NASA spacecraft have made uncrewed flights to **ASTEROIDS**, **COMETS**, and **PLANETS**.

Mercury was the first U.S. program to use piloted spacecraft to explore beyond Earth's atmosphere. Flights were conducted from 1961 through 1963. The first American in suborbital flight, Alan B. Shepard, Jr., proved the flight worthiness of the Mercury spacecraft during his 15-minute, 28-second mission, by which NASA brought the United States into the age of human spaceflight. The following year, John Glenn became the first American to circle Earth, flying parts of the last two orbits manually. The Mercury program concluded in 1963, after astronaut L. Gordon Cooper circled Earth 22 times in 34 hours.

During the flights of the Gemini program, in 1965 and 1966, NASA tested advanced technology to accommodate a two-person crew on more ambitious missions. NASA concentrated on studying the effects on humans and equipment of up to two weeks in space. The agency developed technology to rendezvous and dock with orbiting craft, to maneuver and change their orbit, to work outside the spacecraft, and to perfect methods of reentry and landing. All of these skills were needed for the Apollo program.

Those working on Gemini were able to fulfill all their goals, but the precise instrumentation took longer to develop. The first rendezvous of two spacecraft (*Gemini 6* and 7, December 1965) and the first docking (*Gemini 8*, March 1966) were accomplished, but the highlight of the program was a four-day mission in which astronaut Edward H. White II walked in space (*Gemini 4*, June 1965).

The Apollo program conducted projects from the 1960s through 1975. Although this program was designed to explore the Moon and develop the capability to work in the lunar environment, it also had broader goals, such as establishing the technology to meet national interests in space and achieving U.S. preeminence in space. Under Apollo, in 1969, Neil A. Armstrong was the first human to step on the Moon.

Although the last Apollo mission to the Moon took place in 1972, the Apollo program continued to provide astronautics for other NASA space projects such as Skylab and the joint Apollo-Soyuz Test Project (ASTP), in which NASA designed a docking module that allowed an Apollo spacecraft to dock with a Soviet spacecraft. Docking took place in July 1975.

With the ASTP, NASA fulfilled its defined objectives of international cooperation and peaceful ventures in space. Apollo has also provided technology for launching satellites to study Earth and the Moon and contributed to general astronautics development.

The space shuttle

The space shuttle has been conducting astronautic research since 1981. Improving access to space is this program's goal. The reusable space vehicle and inexpensive launch system allows NASA to conduct space exploration with many different types of equipment, spacecraft, and experiments. NASA's development of the space shuttle enhanced research in the fields of astronomy, astrophysics, atmospheric science, geophysics, life sciences, materials science, gravity, and planetary science, as well as national security.

Altogether, six different orbiters have been built, although one, the *Enterprise*, was used for astronautic research and did not travel to space. In November 1983 the space shuttle orbiter *Columbia* carried the first non-American astronaut, West German Ulf Merbold, as well as *Spacelab 1*, a laboratory designed to conduct scientific research in space. In February 1994 Russian cosmonaut Sergei K. Krikalev flew with American astronauts on the orbiter *Discovery*, inaugurating a new program of cooperation in space between Russia and the United States. In April 1990 *Discovery* deployed the Hubble Space Telescope, which has made many important astronomical discoveries (see TELESCOPES).

Another orbiter, *Endeavour*, carried astronauts back to the orbiting Hubble Space Telescope in December 1993 to perform routine maintenance that would upgrade and extend the life of the observatory. *Endeavour* then reboosted the telescope into its 370-mile (595 km) circular orbit. In 1995 the shuttle program conducted its first in a series of dockings with the Russian *Mir* space station, another cooperative effort between the United States and Russia.

The shuttle has not operated without mishaps, however. In January 1986 *Challenger* exploded during launch from Kennedy Space Center following a filaure in its rocket boosters. Public criticism following the *Challenger* disaster prompted substantial reforms in NASA's management, safety program, and human spaceflight procedures. After being grounded for more than two years, the shuttles began flying again in 1988. However, in February 2003 the shuttle *Columbia* disintegrated during reentry after its wing was damaged by flying debris during liftoff. The remaining shuttles were once more grounded until major modifications to the protective tiles and other systems could be made. The shuttle program is now expected to end around 2010.

The explosion that destroyed the space shuttle* Challenger *on January 28, 1986, occurred 73 seconds after liftoff and killed all seven crew members.

SPACE SHUTTLES

Six different space shuttle orbiters have been built: *Atlantis* (shown to the right), *Challenger, Columbia, Discovery, Endeavour*, and *Enterprise*, which was used for development tests, such as approach and landing, but was not built for travel into space. After the *Challenger* and *Columbia* disasters, the three remaining shuttles with the capability to fly in space have required major modifications to overcome the problems that led to *Columbia's* destruction.The orbiters all had have the same basic design, but there are minor differences, including how long a mission they can conduct and how much they weigh. As NASA began conducting missions to Russia's *Mir* space station, they modified some of the orbiters to be able to dock with the station. The orbiter is propelled into space by strap-on booster rockets and its own engines. When the boosters are jettisoned, the orbiter continues, using only its own liquid-fueled engines. The astronauts operate the orbiter in space like a spacecraft, but after firing its engines to slow it for reentry, they fly it through the atmosphere like an aircraft, landing in Florida or California.

The space shuttle system consists of three parts: the orbiter, the external tank, and twin rocket boosters. The orbiter, which resembles a large airplane with a 60-foot by 15-foot (18 m by 4.6 m) cargo bay, carries a crew of five to eight. The 154-foot- (47 m) tall external tank stores more than 520,000 gallons (2 million liters) of liquid hydrogen and liquid oxygen propellants. This tank serves the shuttle's main engines. At launch, the three engines on the orbiter, fed by the external tank, produce a combined thrust of nearly 1.2 million pounds (5.8 million newtons; N). Once the main engines are shut down, the external tank is jettisoned and burns up in Earth's atmosphere. Recoverable twin solid rocket boosters, strapped to the external tank, generate more than 6 million pounds (29 million N) of thrust at liftoff. These boosters are jettisoned two minutes into flight.

The shuttle can transport people, materials, equipment, and other spacecraft (known as the shuttle's payload) into orbit and enable astronauts to leave the vehicle to work in space, on or in other orbiting spacecraft. The flight frequency is usually around five or six per year. Since its first launch in 1981, the shuttle has carried more than 1,000 payloads to orbit, retrieved and repaired satellites and carried crews to *Mir* and the International Space Station. The space shuttle has also carried experiments and other cargo for NASA and the U.S. Department of Defense, as well as for private, commercial, foreign, and educational organizations.

A CLOSER LOOK

NASA and aeronautics

NASA has also made progress in aeronautics, the study of air flight. NASA research was involved in the development of the forward-swept-wing X-29 aircraft. Similarly low-speed prop-fan engines might power the airliners of the future. NASA also sponsors research into advanced propulsion systems, such as scramjets. The agency also conducts research into the effects of aeronautics advances on humans. Besides this aeronautics technology, NASA's continuing research includes related projects on controlling aircraft pollution and reducing engine noise.

NASA management

Although issues of U.S. and Soviet competition for global leadership in space dominated the origins of NASA, since 1958 the agency's work has stimulated an awareness of the fragile existence of Earth within the universe and fostered a spirit of international cooperation. Astronautics research and technology are being used for a variety of studies focusing on both Earth and space. Most of these can be understood by looking at the way in which NASA is organized.

NASA is divided into managerial divisions that coincide with current projects. Because of the diversity in NASA's goals and objectives, these managerial divisions, no matter what project they are associated with, all focus research and technology development into two strategic areas: spaceflight, which includes the space shuttle and International Space Station; and exploration systems, which include plans to send humans back to the Moon and on to Mars.

Spaceflight

Human exploration and utilization of space is undertaken by the office of spaceflight. Multinational crews are sent on missions to achieve routine space travel, to conduct scientific research, and to enrich life on Earth by means of people living and working in space.

Several centers for the commercial development of space are maintained at universities and research institutions across the United States, sponsoring experiments carried aboard the space shuttle or space station and operated by astronauts or ground controllers. For many years Spacelab was used as a commercial space research laboratory. Today, these experiments are conducted in the Destiny laboratory on the International Space Station (see SPACE SCIENCE).

Mission to planet Earth

The total Earth environment is being studied by NASA's Earth science office, while processes and instruments for long-term environment and climate monitoring and prediction are being developed. In this mission, NASA is trying to increase scientific understanding of Earth as an integrated environmental system and uncover its vulnerability to natural changes and human influence; to characterize the entire Earth environmental and climate system and make the data widely available; to contribute to wise and timely national and international policy; and to foster the development of an informed and environmentally aware public.

Earth's land, vegetation, and water have been studied since 1972 through LANDSAT (NASA's first Earth resources monitoring satellite). As scientists' understandings of the diversity and complexity of Earth's atmosphere, oceans, and land matures and as computers and other tools for studying these systems become more sophisticated, the global nature of the LANDSAT program is being replaced by more specifically focused satellites, such as NASA's Upper Atmosphere Research Satellite, to study ozone, ocean biology, wave height, and ice-covered regions. NASA's Earth Observing System (EOS) is a series of satellites designed to make a wide range of measurements in a coordinated program to treat Earth as a complex, global system of atmosphere, water, land, and life (see ATMOSPHERE; BIOSPHERE).

Space science

The office of space science seeks to answer fundamental questions such as "What is the origin of the universe?" and "Is there life beyond Earth?" One of the main areas of space science investigated by NASA focuses on surveying planets and their moons by using robotic spacecraft. NASA's exploration of the solar system program has included spacecraft encounters with all the planets (except Pluto) and many of their moons (see SPACE EXPLORATION).

In 2004 President George W. Bush, announced a new initiative that would eventually send U.S. astronauts back to the moon and then on to Mars. In order to achieve this objective a new NASA office of exploration systems was established. The fundamental goal of this vision was to advance U.S. scientific, security, and economic interests through a robust space exploration program. In support of this goal, NASA will be required to

- implement a sustained and affordable human and robotic program to explore the solar system and beyond.
- extend human presence across the solar system, starting with a human return to the moon by the year 2020, in preparation for human exploration of mars and other destinations.
- develop the technologies, knowledge, and infrastructures both to explore and to support decisions about the destinations for human exploration.
- promote international and commercial participation in exploration to further U.S. scientific, security, and economic interests.

The office of exploration will conduct a series of robotic preparatory missions to map the Moon's resources, while, at the same time, developing an exploration vehicle that will replace the shuttle and enable humans to land on nearby worlds.

NASA aeronautical research was involved in the development of the forward-swept-wing X-29.

See also: ATMOSPHERE; BIOSPHERE; MARS; MESOSPHERE; SOLAR SYSTEM; SPACE EXPLORATION; SPACE SCIENCE; SPACE SHUTTLES; SPACE STATIONS; TELESCOPES; THERMOSPHERE.

Further reading:

Garber, Stephen J., ed. 2002. *Looking Backward, Looking Forward: Forty Years of U.S. Human Spaceflight.* Washington, D.C.: National Aeronautics and Space Administration, Office of External Relations.
Holden, Henry M. 2004. *Living and Working Aboard the International Space Station.* Berkeley Heights, N.J.: MyReportLinks.com Books.
———. 2004. *The Tragedy of the Space Shuttle Challenger.* Berkeley Heights, N.J.: MyReportLinks.com Books.
Tocci, Salvatore. 2003. *NASA.* New York: Franklin Watts.
NASA: http://www.nasa.gov

NATIONAL WEATHER SERVICE

The National Weather Service is a U.S. government agency that provides information about weather and climate

Tornadoes are highly destructive rotating columns of air characterized by a funnel-shaped cloud. They occur most frequently in the midwestern United States. Wind speeds of 230 miles per hour (370 km/h) have been observed.

CONNECTIONS

- **SATELLITE** and **RADIO AND RADAR** technology is used to forecast changes in **CLIMATE**.

- Meteorologists are sometimes able to predict severe weather conditions such as **HURRICANES** and **FLOODS**.

The National Weather Service (NWS), a branch of NOAA (National Oceanagraphic and Atmospheric Administration), is a U.S. government agency responsible for providing the weather information necessary to protect life and property. The NWS regularly issues short-term weather forecasts and longer-term (one month to one year) climate predictions. Severe weather watches and warnings cover tornadoes, severe thunderstorms, hail, damaging winds, hurricanes, river flooding, and blizzards. Weather services are made available to all users, including the public, through observations and models and through the interpretive skills of forecasters.

Monitoring the weather

Among the weather-monitoring tools are the GOES (Geostationary Operational Environmental Satellites), which observe the United States and adjacent ocean areas from 22,300 miles (36,000 km) above the equator (see SATELLITES, ARTIFICIAL). The NWS receives data from two satellites, which provide continuous day and night weather observations that allow forecasters to monitor severe weather events such as hurricanes, thunderstorms, and flash floods (see FLOODS, HURRICANES).

Satellite data complement the thousands of weather observations made daily by government agencies, volunteer citizen observers, ships, planes, automatic weather stations, and radiosondes (balloon-carried instruments that record pressure, temperature, and humidity) launched from points around the world in an increasingly successful effort to answer the basic question: "What's the weather going to be?"

In efforts to answer this question, the NWS provides myriad localized forecasts through its Weather Service Forecast Offices (WSFOs), Weather Service Offices (WSOs), and River Forecast Centers (RFCs). WSFOs perform the bulk of the forecasting functions for statewide areas, and WSOs and other small offices provide local adaptive forecasts. For example, during the month-long fire in Yellowstone National Park in 1988, NWS provided weather information to the firefighting teams.

Local NWS operations are grouped into six regions: Alaska, in Anchorage; Central, in Kansas City, Missouri; Eastern, in Bohemia, New York; Pacific, in Honolulu; Southern, in Fort Worth, Texas; and Western, in Salt Lake City.

Environmental Prediction

Although the local forecast services are the NWS interface with the public, the National Centers for Environmental Prediction (NCEP) are the heart of NWS operations. NCEP provides central meteorological and oceanographic guidance to field forecasters to assist them in issuing forecasts and warnings to the public. NCEP, formerly the National Meteorological Center, analyzes virtually all the meteorological data collected around the globe. This data includes information from a large network of massive automated moored buoys and coastal stations in all seas bordering the United States.

The network, operated by the National Data Buoy Center (NDBC) at Stennis Space Center, Mississippi, provides hourly measurements of wind speed and direction, atmospheric pressure, air and sea surface temperature, as well as information about the conditions at sea to weather forecasters.

CORE FACTS

- The National Weather Service (NWS) aims to provide detailed short-term weather predictions as well as to forecast long-term changes in climate.
- The NWS provides local and national weather forecasts to universities, aviation authorities, and private organizations as well as to the general public.
- The latest satellite and radar technology is used by the NWS as a means of recording precise changes in climate.
- The NWS has recently embarked upon a comprehensive modernization program to improve its ability to detect and warn of severe weather.

NCEP has seven centers: the Storm Prediction Center in Norman, Oklahoma; the Aviation Weather Center in Kansas City, Missouri; the National Hurricane Center and Tropical Prediction Center in Coral Gables, Florida; the Marine Prediction Center formerly in Camp Springs, Maryland, and moved to Monterey, California; the Climate Prediction Center in Camp Springs, Maryland; the Hydrometeorological Prediction Center in Silver Spring, Maryland; and the Environmental Modeling Center and Central Operations Branch in Camp Springs, Maryland.

The NWS has seven basic services that provide a variety of near-real-time weather and flood data. This information serves customers ranging from the general public to universities and private organizations. The NWS formats include products intended for the general public, domestic and international data, a numerical service, a direct connect service, digital facsimile service, and the computerized graphics system that provides weather information to NWS meteorologists.

The modernized weather service

With increased understanding of the atmosphere and modern technology for observing and predicting the weather, the National Weather Service began a major modernization program in 1989 to improve significantly its ability to detect and warn this concerned of severe weather. Detailed atmospheric data are collected on a routine basis using an array of sophisticated weather-observing systems such as geostationary and polar satellites (see SATELLITES), lightning detection networks (see LIGHTNING AND THUNDER), and wind profilers (see WIND). Such technology complements the NWS network of human observers and spotters, experienced forecasters, and improved numerical forecast models to provide an unprecedented amount of information on the structure of the atmosphere. At the same time that observations have improved, techniques for handling data have also improved dramatically. As a result, the potential for more-accurate short-term warnings and weather forecasts has increased markedly in the last few years. For example, improved signal and data processing is reducing ambiguities in interpreting data from new observing technologies, and conceptual and numerical models enhance a forecaster's ability to translate the new observations into more-effective warnings and forecasts.

Three major development efforts helped modernize NWS operations:

1. A communications and forecaster workstation system known as the Advanced Weather Interactive Processing System (AWIPS).

2. An Automated Surface Observing System (ASOS), which continuously provides surface observations of temperature, wind, humidity, pressure, and other atmospheric variables.

3. A new WSR-88D Doppler radar system (known as NEXRAD in its development phase).

AWIPS is an important part of the modernized NWS. It prepares forecasts, warnings, and other products and services. A complementary part of the modernization plan involves the reorganization of all NWS operations. Beginning from a base of 52 Weather Service Forecast Offices (WSFOs), 13 River Forecast Centers (RFCs), and 180 Weather Service Offices (WSOs), the modernized NWS includes 118 Weather Forecast Offices (WFOs) located with the new WSR-88D radars. AWIPS provides the communications and forecast support functions for these offices.

ASOS serves as the primary surface weather observing network in the United States. It has more than doubled the number of full-time surface stations and provides around-the-clock information on meteorological variables such as cloud height, visibility, temperature and dew point temperature, wind direction and speed, and precipitation accumulation.

Doppler radars target the weather

The WSR-88D radars are 10 times more sensitive than any previous weather radar. The WSR-88D uses the Doppler effect to detect motion within storms. (German scientist Christian Johann Doppler [1803–1853] discovered in the 19th century that an object moving toward a detector will compress light, sound, or radio waves, while an object moving away

Weather satellites orbit Earth and monitor meteorological conditions by, for example, transmitting photographs of cloud patterns and making other meteorological observations that aid in weather forecasting.

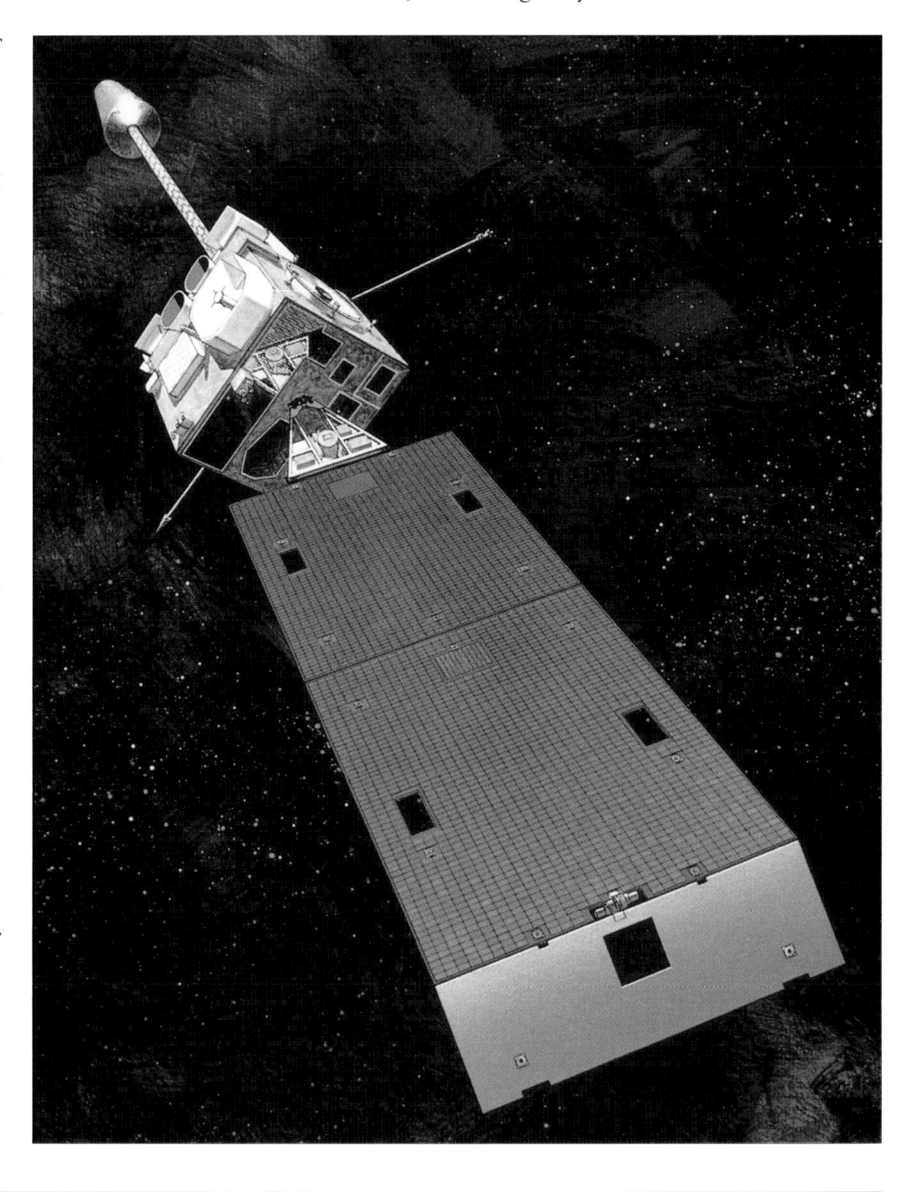

An automatic weather-recording station on the Falkland Islands, 300 miles (483 km) off the Argentine coast, is connected by satellite to the Meteorological Office in Britain.

will elongate them [see DOPPLER EFFECT]. This change is the frequency change heard with passing planes, trains, and sirens.)

The WSR-88D uses microwave signals to gauge motion within a storm; both the speed and the direction of motion inside storms can be calculated. This information allows the radar to identify severe weather conditions. Thus, the radar's Doppler feature allows for greatly increased capability for short-term forecasting, or "nowcasting," of thunderstorms, detection of mesocyclones (vertical columns of rotating air) associated with tornadoes and other severe storm hazards, and warning of damaging winds such as macrobursts and wind shear. Doppler radar is also quite successful in estimating rainfall amounts for use in river and flood forecasting. In comparison with the previous weather radar network, the WSR-88D network covers a much broader area of the United States.

Forecast duties in the modernized WFOs are no longer divided along service boundaries. Instead, forecasters are becoming "nowcasters," responsible for all forecasts for a specific block of time. Forecasters first examine the available data and any guidance offered by NCEP to form a mental picture of the state of the atmosphere. They must then convert their mental pictures to a form usable by the public and other weather customers.

In the modernized weather service, forecasters spend less time preparing weather reports and more time working with weather data to produce better forecasts. Starting with the current forecast database and using a set of highly interactive graphical tools, forecasters can visualize and modify the forecast weather elements. The resulting database is then used to write forecasts. Monitoring and verifying forecasts is also simplified and enhanced by the existence of a forecast database.

History of the National Weather Service

The National Weather Service celebrated its 125th anniversary on February 9, 1996. Both forecasting the weather and taking the observations that make such forecasts possible have changed dramatically in the 125 years since the U.S. government began providing weather services. In 1870 when President Ulysses S. Grant (1822–1885) signed a joint resolution of Congress authorizing the secretary of war to establish a national weather service, weather observations were made by "observer sergeants" of the U.S. Army Signal Service. These were the first systematized, synchronous (collected at the same time) weather observations ever taken in the United States.

The original weather agency operated under the War Department from 1870 to 1891, with its headquarters sited in Washington, D.C., and field offices concentrated mainly east of the Rockies. Very little meteorological science was used then to make weather forecasts. Instead, the weather that occurred at one location was assumed to move into the next area downstream.

From 1891 to 1940 the Weather Bureau was part of the Department of Agriculture. In 1902 Weather Bureau forecasts were sent by wireless telegraphy to ships at sea. The first wireless weather report was received from a ship at sea in 1905. Two years later, the daily exchange of weather observations with Russia and eastern Asia was inaugurated.

In 1910 the Weather Bureau began issuing weekly outlooks to aid agricultural planning. Soon thereafter, weather forecasters began using more sophisticated observational methods, including kites, to measure temperature, relative humidity, and winds in the upper atmosphere.

Realizing that the Weather Bureau played an important role for the aviation community, and therefore for commerce, in 1940 President Franklin D. Roosevelt (1882–1945) transferred the Weather Bureau to the Department of Commerce. In 1942 a Central Analysis Center was created to prepare and distribute information about the upper atmosphere. It was a forerunner of the National Meteorological Center. During the late 1940s the military gave the

Weather Bureau a new and valuable tool—25 surplus radars—thereby launching the first network of weather surveillance radars.

Computer technology in the 1950s paved the way for the formulation of complex mathematical weather models, the result being in a significant increase in forecast accuracy. In 1960 the world's first weather satellite (*TIROS I*) was launched. In 1970 the name of the Weather Bureau was changed to the National Weather Service, and the agency became a component of the Commerce Department's National Oceanic and Atmospheric Administration (see NOAA).

Today the National Weather Service is at the brink of a meteorological revolution. Advances in satellites, radar, sophisticated information processing and communication systems, automated weather observing systems, and superspeed computers are the centerpieces of the modernization that will result in more timely and precise severe weather and flood warnings for the nation.

J. DENNETT

See also: ATMOSPHERE; DOPPLER EFFECT; FUJITA, TED; GLOBAL WARMING; MARINE EXPLORATION; MESOSPHERE; METEOROLOGY; NOAA; OCEANS AND OCEANOGRAPHY; PALEOCLIMATOLOGY; STRATOSPHERE; THERMOSPHERE; TORNADOES; TROPOSPHERE; WAVES, TIDAL.

Further reading:

Ahrens, C. Donald. 2003. *Meteorology Today: An Introduction to Weather, Climate, and the Environment*. Pacific Grove, Calif.: Thomson/Brooks/Cole.

Hughes, Monica. 2004. *Weather Patterns*. Chicago: Heinemann Library.

Sievert, Terri. 2005. *Weather Forecasting*. Mankato, Minn.: Capstone Press.

Wills, Susan, and Steven Wills. 2004. *Meteorology: Predicting the Weather*. Minneapolis: Oliver Press.

National Oceanic and Atmospheric Administration: http://www.noaa.gov

U.S. National Weather Service: http://www.nws.noaa.gov

FORECASTS CAN SAVE LIVES

The weather frequently makes news. Weather forecasts make the news also. One of the most famous weather forecasts was made on Palm Sunday in 1965. The Weather Bureau warned several hours in advance of 33 of the 37 tornadoes that ravaged the Midwest, killing 266 people and injuring 3,261 others. The loss of life and injuries were attributed to people ignoring the timely warnings and continuing to enjoy their Sunday activities.

Sometimes, forecasts are not quite right. Another well-known weather forecast occurred in 1900 for Galveston Island off the coast of Texas. Forecasters rightly warned of an inbound September hurricane. However, they failed to predict an accompanying tidal surge. On September 8, more than 6,000 people died in the storm waters that washed over the low-lying island. It was the worst natural disaster in U.S. history.

On the other hand, new understanding of the unpredictable nature of so-called nonlinear systems, such as the atmosphere (see ATMOSPHERE), suggests that exact forecasts of weather conditions at particular places may never be possible much more than several weeks in the future. As one famous American meteorologist put it, "As odd as it sounds, we may find it easier to control the weather than to forecast it."

SCIENCE AND SOCIETY

This control station for meteorological research into a thunderstorm is at the NWS field center, Cape Canaveral, Florida.

NATURAL GAS

Natural gas consists of gaseous hydrocarbons, which occur alone or in association with oil

Often, when small amounts of natural gas occur in conjunction with oil, the gas is burned off as a gas flare, such as the one pictured above at an oil-production wellhead in Russia.

Natural gas is a hydrocarbon found underground in porous rocks, often in association with deposits of crude petroleum or coal seams (see HYDROCARBONS). Natural gas is used as a fuel to heat buildings and generate electricity and, more recently, as a fuel in transportation systems. It is also used as a raw material for chemical production.

The primary component of natural gas is methane (CH_4), which generally makes up 80 to 98 percent by volume (see ORGANIC CHEMISTRY). Other substances include propane, ethane, butane, and pentane, which are higher-molecular-weight hydrocarbons. Natural gas can also contain smaller amounts of nitrogen, carbon dioxide, and helium. In contrast to methane, which is known as dry gas or marsh gas, propane, butane, and pentane are liquid at moderate conditions of temperature and pressure.

Chemically, methane is the simplest of the hydrocarbons. Since it also has the highest ratio of hydrogen to carbon, it will burn cleanly in air (see FUNCTIONAL GROUPS). These chemical characteristics increase the attractiveness of natural gas as a fuel, since the energy supplied comes with few harmful by-products (see POLLUTION).

CONNECTIONS

- The underlying **GEOLOGY** of an area is important in determining the location of **OIL** and natural gas deposits.
- Natural gas is becoming increasingly important as a global **ENERGY** source.

CORE FACTS

- The principal component of natural gas is methane (CH_4).
- Chemically, methane is the simplest of the hydrocarbons and is characterized by the highest ratio of hydrogen atoms to carbon atoms.
- Natural gas occurs in the pores of certain rocks trapped within Earth's crust. It is captured and brought to the surface through drilling.
- Exploitation of natural gas as a fuel depends on vast systems of distribution pipelines to carry the gas from the point of extraction to the point of use.

Origin

Like oil, natural gas is a fossil fuel produced from the decomposition of microscopic animals. Most organic material decomposes completely through chemical and microbial activity. Under waterlogged conditions, however, the incomplete oxidation of the original organic materials, allows it to decompose much more gradually. Over time, these conditions lead to deposits of oil and gas becoming trapped in rocks within Earth's crust.

It is thought that most oil and gas resulted from decomposition in shallow marine environments, such as are found on continental shelves. For two reasons, however, many of the world's largest oil deposits are now found under land. First, many oil-rich land areas were originally areas covered by the sea. For example, the oil-rich Middle East once formed part of a continental shelf. The second reason is that oil and gas migrate under pressure, (caused by overlying strata), from source rocks to reservoir rocks (porous rocks in which large quantities of oil and gas accumulate).

The exploited reservoir rocks of the world's oil fields contain much greater quantities of oil and gas than could have been originally formed in these particular spots. Geologists distinguish between two types of migration: primary migration, which occurs out of very small openings in the source rocks, and secondary migration, which occurs through the reservoir rocks to areas called traps from which oil and gas cannot escape. Deposits of natural gas and oil are commercially exploitable when they accumulate in sufficiently large amounts to repay the costs of extraction.

Uses of natural gas

Although the Chinese used geologically occurring methane 2,000 years ago, natural gas has become a major energy source only in this century.

Natural gas is distinguished from the types of manufactured fuel gas that preceded the major discoveries of the natural deposits. Coal gas was manufactured from coal, for example, and then used for heating (see COAL AND COAL GAS). Compared to coal gas, natural gas has a lower flame speed (burns more slowly) and larger flame size but a calorific value (heating power) of more than twice as much.

The initial recovery of natural gas was in conjunction with oil drilling rather than an activity in its own right. The widespread exploitation of natural gas as a fuel began in the 1930s, following the construction of vast systems of pipelines. Unlike oil, which can be transported easily in barrels and tankers, gas requires distribution networks to connect the extraction site with the user. Once in place, however, its ease of use far surpasses that of oil. Natural gas requires very little treatment before being useful as a fuel in stoves, boilers, or power plants.

Natural gas has become an extremely important fuel source and is probably destined to become even more vital in coming years, in part because its relative environmental benefits compared with the pollution threats posed by oil and coal. Perhaps more important has been the discovery of new reserves and the related development of new technologies that enable greater efficiency in the use of natural gas.

Exploration and production

When geologists search for new oil and gas fields, they begin by studying maps of surface conditions. When an area looks like a reasonable prospect, one possessing likely looking large-scale rock structures that might contain oil or gas traps, they investigate the underlying geological conditions.

One standard technique is seismic reflection profiling, which involves the use of vibrations. It initially involved setting off a series of controlled explosions to produce seismic waves similar to those found in earthquakes. The practice has evolved into a less ecologically disruptive method, one involving generating seismic waves by thumping the ground. The logic behind this technique is that the time it takes for the various reflected waves to return yields insight into the characteristics of the underlying geology. When these techniques show the presence of suitable oil or gas traps and other structural features, the exploratory drilling phase begins.

Bringing the gas to the surface generally requires application of a force to extract it from the pores of the reservoir rocks. Methods include fracturing the reservoir rocks with explosives and using pressure by pumping water or air into the deposits to force the gas up.

Major world producers of natural gas

The major producers of natural gas in 1990 were the former Soviet Union and the United States, which together accounted for more than 60 percent of global production. In the late 1970s the lifting of the price controls on gas, by making new supplies more profitable, was largely responsible for the surge in United States production.

The distribution of global production is likely to shift in the future. As far as proven reserves go, the United States is less important, with under 4 percent of total global reserves. Proven reserves are those reserves that can be recovered under existing economic and operating conditions. Russia and Ukraine, with 33 percent of proven reserves, and Iran, with 14.3 percent, are likely to dominate the future natural gas production scene.

Recent years have seen a marked expansion in the production and use of natural gas around the world. Indeed, there are some energy analysts who maintain that it could ultimately provide 90 percent of the global energy demand.

S. FENNELL

FAULT TRAP

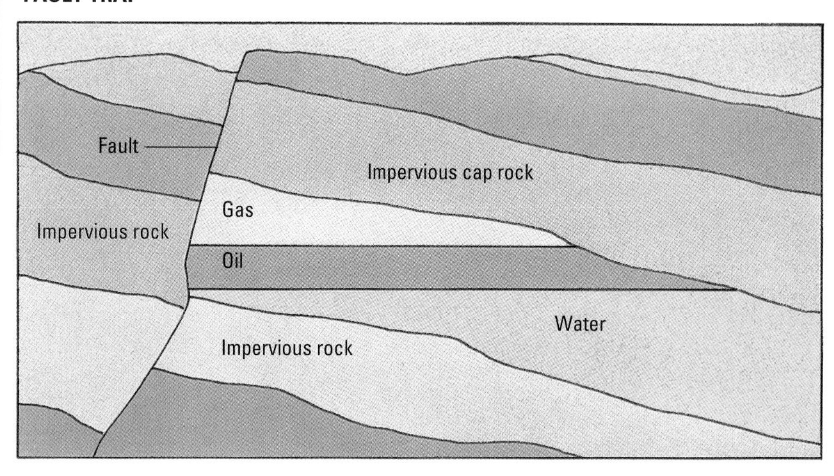

ANTICLINE TRAP

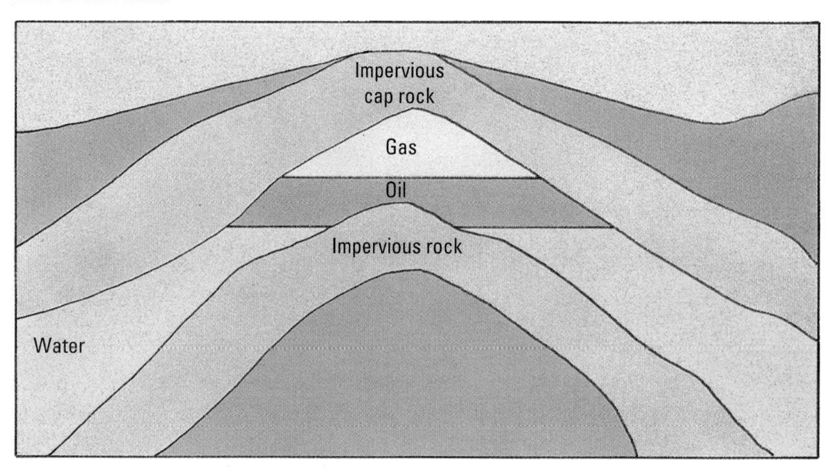

SALT DOME

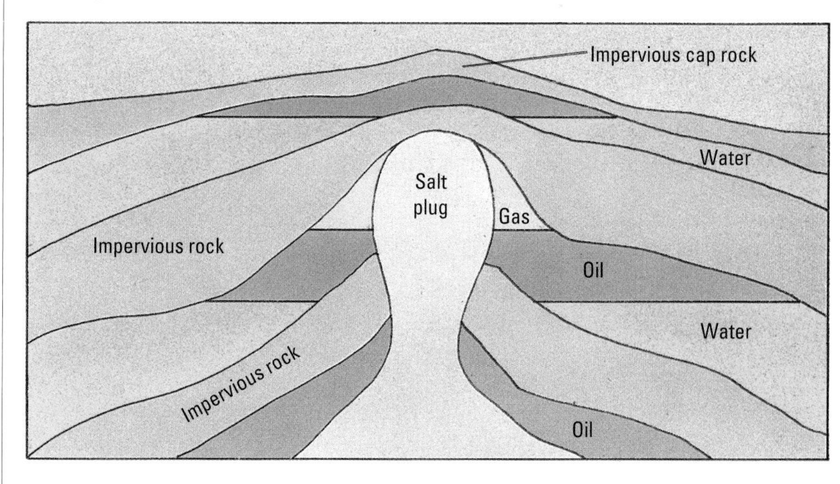

Three examples of the rock structure in Earth's crust that might lead to the formation of oil and gas traps. In each case, the oil and gas cease their upward migration when they reach impervious rock.

See also: COAL AND COAL GAS; FUNCTIONAL GROUPS; HEAT; HYDROCARBONS; MINING AND PROSPECTING; ORGANIC CHEMISTRY; POLLUTION.

Further reading:

Busby, R. L. ed. 1999. *Natural Gas in Nontechnical Language*. Des Plaines, Ill.: Institute of Gas Technology.

NAVIGATION

Navigation is the science of determining location and plotting a course to a destination

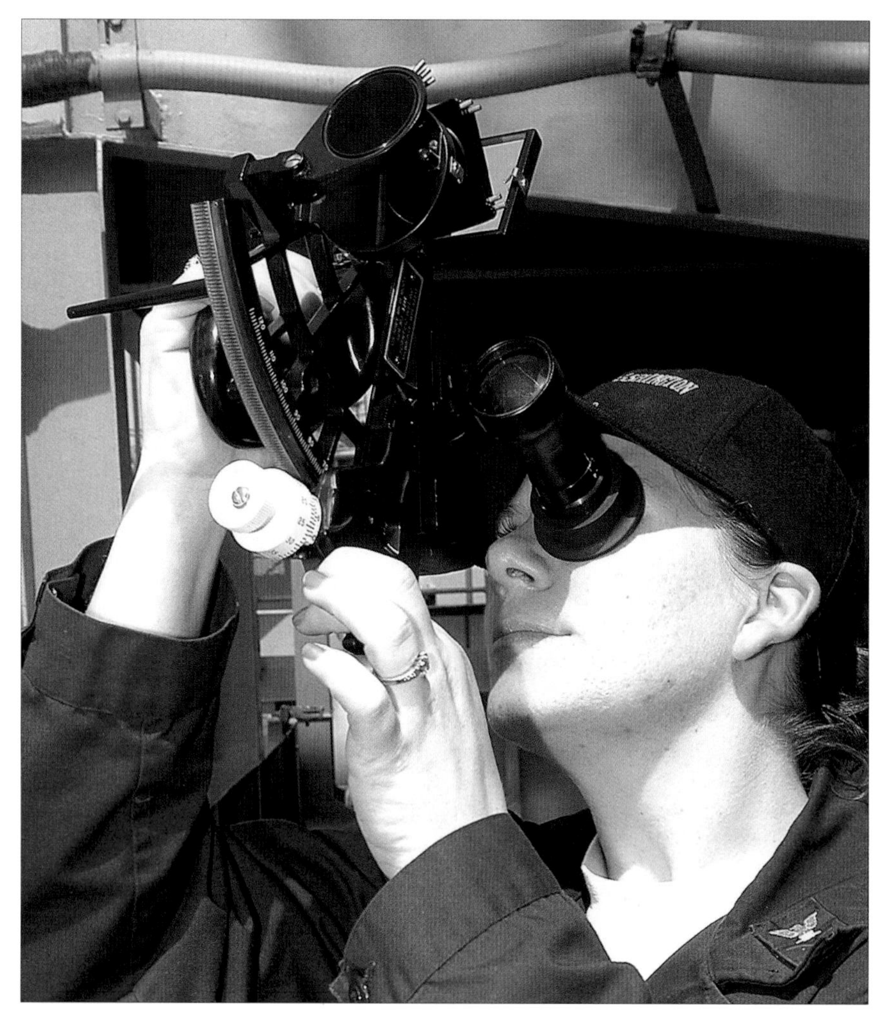

A U.S. Navy sailor uses a sextant to measure the position of the Sun above the horizon. This measurement will allow her to calculate the ship's latitude—position north or south of the equator.

The word *navigation* comes from *navis*, the Latin word for "ship;" ships were the first means of transport to travel over long distances, and therefore spurred the development of navigation. Put simply, navigation is the craft of finding one's location on the Earth, and then plotting a course to a destination. To do this, navigators and cartographers divide the surface of the Earth with lines of longitude—running around the planet and passing through both poles—and lines of latitude—those running parallel to the equator (see MAPS AND MAPPING).

In this way, any point on Earth can be given coordinates in longitude and latitude, measured in degrees. There are 360 degrees in a circle, so longitude can be measured in degrees east or west of any known position on the surface of Earth.

By convention, longitude today is measured relative to the Greenwich meridian, the line of longitude passing through the Royal Greenwich Observatory in London, England. Latitude, meanwhile, is measured in degrees north or south of the equator, up to a maximum of 90 degrees. For further accuracy, 1 degree (1°) can be divided into sixty minutes (60'), and in turn, each minute divided into sixty seconds (60").

Latitude was fairly easy to find, at least in theory, even for ancient sailors, by measuring the position of heavenly bodies in the sky (see below). However, an accurate means of finding longitude at sea was possible only after the development of accurate clocks in the 18th century. Since that time, technology has created new and faster means of transportation and opened up a whole new realm to navigate—the air. The science of navigation has kept pace with the technology of transport. Today people can find their position on Earth to within a few meters using a simple handheld electronic device.

Coastal navigation

The earliest navigators sailed in coastal waters and rarely strayed into the open oceans, but they still needed a means to accurately plot their location and course to find safe harbors and avoid being driven onto rocks or submerged sandbanks.

The simplest method of coastal navigation was simply to look for landmarks along the shore that might act as guides. A more advanced version of this method was to look for an alignment of two landmarks. Such alignments could be extremely accurate because they could be seen only if a boat lay along a specific line, but useful alignments were unlikely to happen by chance. However, they were easy to create artificially, using one or more beacons, moored buoys, or other humanmade landmarks. Such markers are still widely used in coastal navigation today. They indicate, for example, deepwater channels that remain safely navigable at low tides.

The next step is to find a vessel's position when there is no convenient alignment. In practice a sailor would use a map of the coast and a device for taking a bearing (measuring the direction) of known landmarks. A single bearing places a vessel on a line, and a second fixes its position along that line.

CONNECTIONS

- People can navigate using compasses because they always point to the **MAGNETIC POLES**.
- Being able to measure **TIME** accurately is essential for navigation.

CORE FACTS

- Navigators use a variety of skills to find a vessel's position and plot a course to a destination.
- Latitude can be found by measuring the positions of the Sun and stars in the sky, but longitude measurement requires an accurate clock.
- On short journeys close to land, navigators can find their location and course by observing fixed landmarks.
- Compasses and ships' logs allow a navigator to measure their direction and speed. Advanced versions of these instruments allow aircraft navigators to find the same information.
- Radio navigation transformed travel by allowing extremely accurate location finding. The latest development in this field has been the GPS satellite navigation system.

Although the techniques of the coastal navigator, or pilot, have been refined over centuries, they are useful only when in sight of land. Far out to sea, navigators must rely on other clues to their position, and for many centuries these clues came principally from the sky.

Navigating by the stars

Even before Earth's spherical shape was known, seafarers were taking advantage of its effects. For example, the elevation of the Sun at midday (when it reaches its highest point in the sky) varies depending on the observer's latitude (see DAY; MIDNIGHT SUN). If the Sun is overhead at the equator, it will lie close to the horizon for someone near the North Pole.

Ancient navigators realized that they could, in theory, find their position on Earth by measuring the Sun's midday elevation, but there were complications. Because of the tilt of Earth's axis, the Sun's daily path across the sky drifts north and then south again in the course of a year (see SEASONS). However, once these variations were accurately understood, vessels going on long voyages could carry almanac tables, which gave the Sun's elevation at midday throughout the year.

Of course, the Sun was no help at night, but fortunately there are other aids to navigation in the night sky. The Earth's daily spin makes the sky seem to rotate around two fixed points—the north and south celestial poles. By a lucky coincidence, the north celestial pole is marked by a bright and easily identified star known as Polaris, or the Pole Star. Because it lies directly overhead at the North Pole, the elevation of Polaris above the northern horizon is equal to an observer's latitude.

The Southern Hemisphere, has is no bright star at the celestial pole, although its direction can be found by reference to the stars around it. It can still act as a direction indicator, but it is useless for measuring latitude. However, the highly sophisticated Polynesian navigators solved this problem (see OCEANIA). Aware that changing latitude affects the positions of all the stars, they created accurate lists of which stars passed directly overhead at which latitudes. This information was passed on in songs.

NAVIGATIONAL INSTRUMENTS

Traditional navigational instruments fall into two basic categories—those devices, such as the sextant, that measure the position of objects in the sky and are used for finding latitude, and those, such as compasses, that are used for finding directions.

Finding latitude

The first instrument used for measuring the elevation of the Sun and stars was the astrolabe. This astronomical device consists of a disk of metal with a 360-degree circle marked around the outside and a sighting device attached to the center. With the astrolabe hanging vertically, the user looked along the sight to the target object and simply read off its elevation from the marks along the edge.

ANCIENT MARINERS

Unfortunately there is no way of telling precisely how prehistoric sailors navigated. They left no records for historians to study. However, the basic principles of coastal navigation were probably in use long before they were first written down. The Phoenicians of Tyre (now in Lebanon) were probably the greatest seafarers of the ancient Mediterranean, and they and the ancient Greeks wrote detailed accounts of their travels in peripluses—guide books with detailed navigational instructions that allowed any suitably equipped sailor to follow the path of the original explorer. By medieval times the periplus had evolved into the portolan, or pilot book—a combination of charts and directions that was widely used until the arrival of the first accurate maps.

While the Greeks and Phoenicians explored from the Mediterranean Sea down the coast of Africa and as far north as Iceland, they mostly stayed close to the coasts and rarely ventured far out into the open ocean. The sailors of other cultures, however, traveled much farther. The Polynesians undertook great journeys to find and colonize tiny islands across the South Pacific, finally reaching New Zealand around 1000 C.E. The Vikings, meanwhile, famously crossed the Atlantic from Scandinavia and established temporary European colonies in Greenland and North America centuries before Columbus.

These ancient seafarers developed a sophisticated understanding of the sea, nature, and the stars in order to make their epic voyagers. Both the Polynesians and the Vikings understood the significance of currents and tidal patterns caused by distant landmasses, and they also watched and learned from the flight patterns of seabirds.

HISTORY OF SCIENCE

The quadrant, a simplified version of the astrolabe, used the same principle but with only a quarter-circle. While these devices were very accurate on land, getting a good reading from the pitching deck of a ship was a very different matter.

The cross staff was a simple device developed in the 15th century that addressed this problem for the first time. It consisted of a long horizontal rod and a short vertical one that slid back and forth along it. The sailor looked along the horizontal rod so it lined up with the horizon and then moved the vertical rod in or out until its tip aligned with the object being measured. A scale along the horizontal rod then revealed the elevation measured.

The crossstaff was widely used for several centuries before better replacements came along in the form of the octant and the sextant. These sophisticated instruments use a series of adjustable mirrors and an eyepiece to let the user view both the object being sighted and the horizon, overlaid on one another in the same field of view. By making minute adjustments, it is possible to place the Sun or star's image directly on the horizon and get a far more accurate reading than was possible before. Such instruments continued to be widely used until the rise of satellite navigation in the late 20th century.

Compasses

The magnetic properties of certain minerals were well known in the ancient world, but it was the Chinese who first put them into good use. Early compasses were made from lodestone (a magnetic iron ore) fashioned in a ladle shape so that they would

Compasses make use of the fact that a freely moving magnet will always point in the same direction—toward the magnetic north pole.

spin freely when placed on a board and line up with Earth's magnetic field (see IRON AND STEEL; MAGNETISM; ORES). By the late Middle Ages, increased contact between China and Europe meant that compasses were in wide use among European navigators.

However, a magnetic compass has shortcomings that its users must take account of. Earth's magnetic poles are some distance from its poles of rotation, so magnetic north will lie in a different direction from geographical (true) north (see MAGNETIC POLES). The difference between the two is called declination and varies according to the compass's position on Earth. Declination can be plotted on a map and taken into account, but another problem is compass deviation caused by the magnetic fields of metal objects on board ship. The solution to this problem is to isolate the compass as much as possible and to calibrate it before leaving port.

Despite these drawbacks, compasses soon became sophisticated and accurate instruments. A typical ship's compass consisted of a needle on a delicate pivot isolated in a sealed bulb and mounted on a stand called a binnacle. A binnacle was able to rotate in all axes, so whichever way the ship tilted, the compass always remained horizontal. It was even possible to fit tiny magnets inside the mount to compensate for any errors caused by the rest of the ship.

Large modern compasses are no longer magnetic, and ships' captains no longer need to worry about deviation and declination. Instead, they rely on the principle of the gyroscope (see MECHANICS). When a spinning top with a heavy disk around its center is spun rapidly, forces generated by Earth's rotation cause the top's axis of rotation to line up with Earth's own. Electrically powered gyroscopes in sealed units can give far more precise and consistent measurements of direction than magnetic compasses and have the added benefit that they naturally align with geographical, rather than magnetic, north. They are widely used in aircraft as well as onboard ships.

Finding longitude

Accurate timekeeping is another surprisingly vital element of navigation, and the lack of a clock capable of keeping good time for weeks or months at sea was a life-threatening problem for sailors before the 18th century. Although the elevation of the Sun and stars makes it easy to find latitude on Earth's surface, it offers no clue to longitude.

Wherever an observer stands on Earth along the same line of latitude, the Sun will appear to rise and set at the same times and cross the meridian (the imaginary north-south line across the sky) at noon local time. Without a way to directly measure longitude, sailors were reduced to guesswork based on their estimated speed and how long they thought they had traveled. In 1707 the final straw came when a British fleet was wrecked in fog off the Scilly Isles at the tip of southwestern England. The admiral had thought the ships were safely in the mid-Atlantic. The British Admiralty offered a substantial prize to anyone who could come up with a reliable method of finding longitude.

In theory, it was simply a case of finding out the time difference between local noon and noon at a known location. Since Earth rotates once every 24 hours, local noon occurs an hour later for every 15 degrees farther west one travels. The problem, however, lay in finding the time at the home port when a vessel experienced local noon. Most people thought that astronomy was the answer. The phases of the Moon and the positions of the moons of Jupiter were both touted as possible standards of time. The prize was eventually claimed by English clockmaker John Harrison (1693–1776). He spent decades developing the first truly accurate ship's clock, or chronometer, able to keep accurate time on the pitching deck of a ship regardless of damp, corrosion, and temperature variations.

Measuring speed

Finding a vessel's location solves many of the problems of navigation but not all of them. A compass allows a vessel to sail in a particular direction, but without some idea of the speed it is traveling, a captain will have to perform laborious longitude and latitude calculations every time he wants to check his exact location. The process of quickly working out an estimated location on the basis of a vessel's speed, its direction, and its earlier position is known as dead reckoning.

Devices for measuring a ship's speed were developed in ancient times. The simplest was the ship's log—a floating piece of wood thrown out behind the boat and attached to it by a long rope. The rate at which the rope spun out as the boat left the log behind could be measured using knots tied at even intervals along its length. Even today, a ship's speed is measured in knots. One knot is 1 nautical mile per hour. A nautical mile is 1.15 miles (1.85 km) or 1" of longitude.

Ships logs in use today are far more advanced than this: patent logs are dragged behind the ship and contain a spinning propeller that measures the speed of water pushed through them, as well as free-spinning propellers and even pressure sensors attached directly to a ship's hull.

Modern navigation

The arrival of radio in the late 19th century transformed navigation forever. It was put to use almost immediately for transmitting time signals, which allowed ships to correct their chronometers on long voyages and thus to fix their positions with newfound accuracy. The real revolution came with the development of navigational radio beacons.

Radio compasses and ranges

An early application of radio beacons was radio direction finding, which relies on the fact that an antenna in the shape of a loop will pick up the strongest signal when its open face points into the incoming radio waves. By finding the bearings of two distant radio beacons—identified by unique signal frequencies—a vessel's position could be worked out with great accuracy. The same technique was soon applied in aircraft and automated in the form of radio compasses that automatically turned to face the strongest signals.

As the range of radio signals increased, another form of radio navigation was developed. Radio ranges consist of a series of transmitters, each broadcasting radio signals of varying frequency. There are various configurations, but the fundamental principle is to measure the way the signals overlay. The signals will reinforce or interfere with each other depending on the receiver's distance from the transmitters, as peaks or troughs of the radio waves pass at different times, and this difference can be used to work out the receiver's position (see DIFFRACTION AND INTERFERENCE). For example, the simplest radio ranges worked by sending out two identical signals precisely a half cycle out of phase with each other. When the signals combined into a continuous tone, the receiver was midway between them. As technology advanced and it became possible to measure the phase difference between signals more accurately, the idea was extended to the omnirange system, in which a vessel (or aircraft) could work out its position anywhere it could receive a signal.

The next significant advance was the loran (long-range navigation) system, which relies on very precise measurement of the time between two signal pulses and the fact that radio waves cannot travel faster than the speed of light (189,291 miles per second; 299,792 km/s). A loran system consists of a series of master transmitters, each paired with a distant slave transmitter. The master transmits short pulses on a unique frequency. When the slave transmitter receives a pulse from its master, it retransmits it immediately. The loran receiver therefore picks up the pulse from the master station, followed a fraction of a second later by its echo from the slave. The delay between the two signals places the receiver on a specific curve of possible locations. By retuning to another loran station, the operator can easily fix his or her precise location.

PLOTTING A COURSE

With a reasonable idea of both direction and speed, it is possible to plot a course from one place to another. However, several further problems arise, not least of which is how to translate the curved surface of the Earth onto the flat plain of a chart and back again.

This problem frustrated sailors and mapmakers alike for centuries, until Flemish cartographer Gerardus Mercator devised the map projection that bears his name in 1569. Mercator's projection turns lines of latitude and longitude on the surface of the Earth into parallel straight lines on the map and thus makes plotting a straight-line course on a constant bearing very easy. However, this straight-line, or rhumb-line, course is not the most direct route between two points on Earth's surface. Because Earth is a sphere, the shortest route between two points on its surface is actually along a "great circle"—a circle, such as a line of longitude or the equator, with a center that passes through the center of the Earth (see GEODESY).

Navigators can use a different map projection, called a gnomonic chart, to work out these courses. The gnomonic projection turns great circle courses into straight lines, but because such a track will always carry a vessel into higher latitudes, perhaps coming dangerously close to polar ice, the course is often broken into two separate great circles.

This great circle track is then transferred onto a Mercator-projection sea chart, where it becomes a curved line drawn using compasses. The chart makes it easy to measure the bearing that should be taken at every point on the course, but because it is not practical to follow a constantly changing course, ships in reality follow a series of short straight-line courses approximately matching the curve of the true great circle track.

A ship's navigator plots a course of a sea chart using compasses.

A CLOSER LOOK

Aerial navigation

The basic principles of radio navigation lie at the heart of modern aerial navigation as well. Radio ranges are used to provide signals that autopilot and instrument landing systems can lock onto, guiding aircraft across the oceans and helping pilots land in poor visibility.

The other major challenge of aerial navigation lies in finding one's altitude and air speed. Both can be worked out using a device called a pitot tube—a horizontal cylindrical tube on the aircraft's hull through which air is driven as the aircraft moves forward. Pressure sensors aligned horizontally and vertically measure the different components of the air pressure and reveal the forward speed. Barometers can also be used for measuring height above sea level. Altitude can also be found by a radar altimeter, which bounces pulses of radio waves onto the ground below and measures the time they take to return (see AIR PRESSURE AND BAROMETERS).

Most aircraft also use an advanced form of dead reckoning called inertial navigation. An inertial navigation system consists of a gyroscopically stabilized platform carrying highly sensitive accelerometers, aligned with the aircraft's three axes of movement (see ACCELERATION). By recording the acceleration experienced in all these directions throughout the flight, the aircraft's navigational computer can work out its precise speed and orientation at any moment. A similar system is used in spacecraft, which have no immediate surroundings against which to measure their movement.

The GPS system allows someone with a handheld receiver to find his or her precise location at any point on Earth. The satellite system was created to guide bombs and missiles but is now also used for civilian purposes.

Satellite navigation

The art of navigation has transformed completely since the beginning of the space age in 1957. Systems for navigating using radio signals from satellites high above Earth were first developed by the United States and the former Soviet Union for military purposes, but the end of the cold war saw the technology, known as GPS (global positioning system) made available to civilians in a portable format.

Today there are two operational GPS systems, the U.S. Navstar satellites and the Russian GLONASS, with another European system under development (see SATELLITES, ARTIFICIAL). All follow the same basic principle—ringing Earth with satellites whose orbits allow at least four to be visible above the horizon from anywhere on Earth at any time. The satellites orbit at an altitude of around 12,625 miles (20,200 km), and each one carries an extremely accurate atomic clock.

A GPS receiver finds its position by picking up a time signal broadcast from the satellites. The unit contains a sophisticated electronic ephemeris, which lets it work out the precise position of all the satellites at any moment. When it picks up the time signal from a specific satellite, it compares it with the time according to its own internal electronic clock. The time difference reveals how long the radio waves traveled to reach the receiver, and therefore how far away the satellite is at that instant. By measuring signals from three satellites, the receiver can plot where its possible positions overlap and place itself on Earth's surface to an accuracy of just a few feet. On board aircraft or in other systems required to calculate altitude as well as latitude and longitude, a fourth satellite signal is included in the calculations.

In order to take account of minute variations in the satellite orbits and errors in the receiver unit clocks, a new ephemeris is uploaded to the satellites each day and transmitted from them to the receiver units to allow them to reset their internal clocks in time with the satellite clocks and stay accurate.

GPS has revolutionized navigation and brought with it unprecedented accuracy—even though the U.S. Navstar signal is deliberately degraded to limit its accuracy for security reasons. Satellite navigation units fitted in cars can now direct drivers from door to door, while handheld units have come to the aid of lost climbers, sailors, and other adventurers.

G. SPARROW

See also: ACCELERATION; AIR PRESSURE AND BAROMETERS; DAY; DIFFRACTION AND INTERFERENCE; GEODESY; GEOGRAPHY; GIS; MAGNETIC POLES; MAPS AND MAPPING; MECHANICS; MIDNIGHT SUN; OCEANIA; SATELLITES, ARTIFICIAL; SEASONS.

Further reading:

Hofmann-Wellenhof, B., K. Legat, and N. Wieser. 2003. *Navigation: Principles of Positioning and Guidance*. New York: Springer.

Sobell, Dava. 1998. *Longitude*. New York: Fourth Estate.

NEBULAS

A nebula is a cloud of interstellar gas and dust that covers a fixed region of space

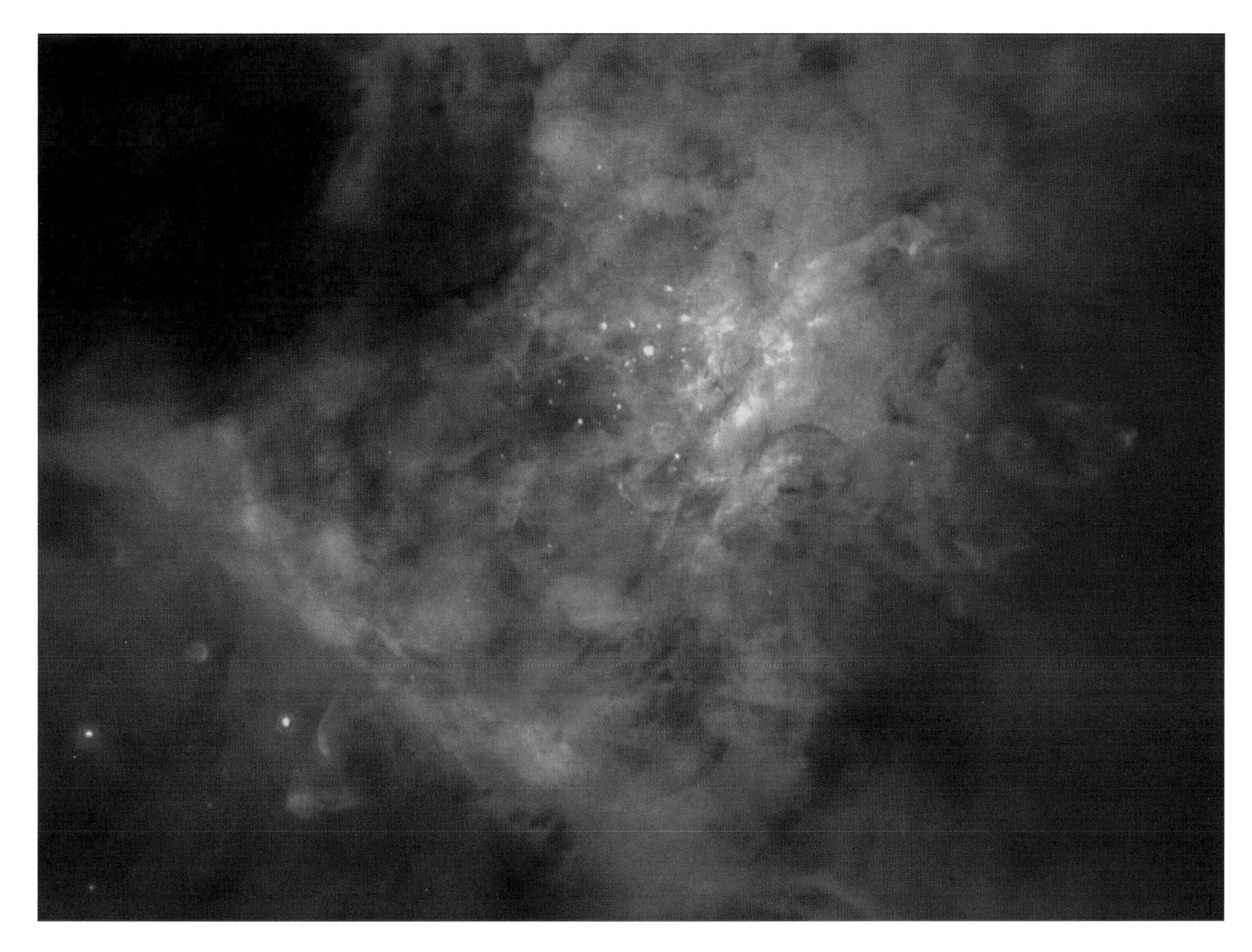

The emission nebula NGC 6334 is located in the constellation of Scorpius.

The word *nebula* (from the Latin word meaning "cloud") was originally used to refer to any fixed fuzzy patch of sky. Early astronomers did not know what a nebula was, and some insisted that a sufficiently powerful telescope would show that any nebula consisted of a number of stars. Indeed, some nebulas turned out to be star clusters or galaxies, but they are no longer referred to as nebulas. Today a nebula is defined as a cloud of interstellar gas and dust that covers a fixed region in space.

There are thousands of nebulas in the Milky Way galaxy (see MILKY WAY), and scientists have developed a variety of systems to classify them. In 1774 French astronomer Charles Messier (1730–1817) published a list of 45 nebulas as a guide for comet hunters who, like himself, could easily confuse a blurry nebula with a comet. His list eventually grew to 103, with each nebula assigned a number. For example, the Ring Nebula in the constellation Lyra is designated M57 because it was number 57 on Messier's list, and the famous Orion Nebula is known as M42 (see the box on page 1033; CONSTELLATIONS). At the time, many astronomers did not understand what a nebula actually was. For this reason, Messier's list includes many celestial objects, such as galaxies and star clusters, that, though blurry in a small telescope, are not nebulas at all (see GALAXIES).

More recently Danish astronomer J. L. E. Dreyer (1852–1926) compiled a larger catalog of nebulas, galaxies, and star clusters—the *New General Catalogue*

CORE FACTS

- The dense mass of gaseous clouds and dust that blocks or scatters the light energy from stars and other bright objects is called a dark nebula.
- Reflection nebulas, often blue, are detected when light reflects from stars toward Earth.
- Ultraviolet radiation causes the atoms in emission nebulas to ionize—light and radio waves are released when these atoms fall back to their stable state.
- As helium fusion causes a mature star to eject its coolest layers into space, a planetary nebula forms.

CONNECTIONS

- The Hubble Space Telescope has revealed that the **GASES** and dust from the Orion Nebula may be a source of new **SOLAR SYSTEMS**.
- Dark nebulas contain a mass of **CLOUDS** that blocks the **LIGHT** from **STARS**.

The Coal Sack Nebula. This dark nebula is found on the border between the constellations of Crux (Southern Cross) and Musca.

of Nebulae and Clusters of Stars (1888, with supplements in 1895 and 1908). In that list, which is more commonly cited by astronomers than the Messier Catalog, the Ring Nebula is known as NGC 6720.

Dark nebulas

A dark nebula is just that—dark (see DARK MATTER). It consists of a dense mass of dust that blocks the light from stars or other celestial objects. Dust grains in the cloud may either absorb some of the light or scatter it from its original trajectory. Therefore, a dark nebula can be detected from Earth only if it lies on a direct line between Earth and a field of stars or other bright objects that the nebula can obscure. In this case, the dark nebula's silhouette will be seen against the bright background. The Horsehead Nebula in Orion is a famous example of a dark nebula.

Emission nebulas

Dark nebulas are detectable because they are dark against a bright background. By contrast, emission nebulas are just the opposite: they are noticed because they are hot and bright. This brightness may occur for several reasons, but in all cases there must be an external source of energy.

One possibility is that the nebula is illuminated by a nearby star or stars. In this phenomenon, ultraviolet radiation (see ULTRAVIOLET RADIATION) from the star causes the atoms in the nebula to ionize (atoms temporarily lose electrons; see IONS AND RADICALS). As the electrons lose some of their energy and combine with positive ions, they emit energy in the form of light and radio waves, which can be detected on Earth. This particular type of emission nebula occurs most commonly near regions of space that have many hot, young stars, which may have formed from part of the nebula. Such masses of these stars can emit enough energy to ionize hydrogen atoms in the nearby nebula, separating each atom's single proton and single electron into an ionized form known as H II (see HYDROGEN). Therefore, these emission nebulas are known as H II regions, and the existence of such a nebula is taken as strong evidence that new stars are being formed nearby. The Orion Nebula is the best-known example of an H II region.

Planetary nebulas are another type of emission nebulas, so named because they have a disk-shaped appearance that resembles a planet. However, they are not solid bodies and are far larger than any planet. The best known planetary nebula is the Ring Nebula,

THE ORION NEBULA

One of the most famous and most exhaustively studied nebulas is the Orion Nebula (shown below), known in the Messier catalog as M42. This nebula is easily seen by the unaided eye in the constellation Orion on any clear winter night in the Northern Hemisphere (see CONSTELLATIONS). To the eye, the nebula looks like a large and somewhat fuzzy star. However, even a small telescope reveals that the object is not a star but a bright, wispy cloud with an irregular shape. It is about 1,500 light-years away from Earth and about 16 light-years in diameter. (One light-year is equal to the distance traveled by light through a vacuum in one year, or 5.88 trillion miles [9.46 trillion km].)

The Orion Nebula is principally an emission nebula: astronomers used spectroscopy to scrutinize the light coming from the nebula and found that the spectrum was mostly dark but also contained bright emission lines, these lines indicating the presence of hot hydrogen, helium, and oxygen in the nebula. The existence of those emission lines was an immediate indication that the nebula was not just reflecting light like a mirror but rather was being heated by some other source and glowing as a result. The source of energy appears to be nearby stars in a group called the Trapezium, a view suggested by the fact that the spectrum of light coming from the nebula includes not only the individual bright emission lines characteristic of rarefied gas but also a weak continuous spectrum, as is characteristic of starlight scattered by dust.

There is much else of interest, however, about the radiation emitted from the Orion Nebula. In 1965, scientists studying its emissions noted something odd: radio astronomers observed that hydroxyl radicals—a bound combination of one hydrogen atom and one oxygen atom (see IONS AND RADICALS)—in the cloud were producing strong radio signals at the microwave frequency of 1665 megahertz (MHz). The scientists deduced that the cloud of molecules in the Orion Nebula acts much like a laser, but in this case, it was called a maser (see LASERS AND MASERS) because the process produced microwaves rather than light. In the Orion Nebula's cloud, hydroxyl radicals apparently absorb infrared energy from starlight, which excites them to a highly energized state. The excited molecules eventually lose this energy, producing mostly 1,665 MHz microwaves in the process, just as the radio astronomers observed. Orion's hydroxyl radical maser was the first to be observed in space, but since then many others have been identified, both in the Orion Nebula and in other nebulas.

The Hubble Space Telescope (see TELESCOPES) has spent much time observing the Orion Nebula, and its photographs have revealed at least 153 protoplanetary disks, which are potential solar systems that are forming out of the nebula's gas and dust. Each of the disks has a young star at its heart, and the material in the disk eventually could coalesce to form planets, just as astronomers believe happened in the solar system.

A CLOSER LOOK

EDWARD BARNARD

Some of the most important early observations of nebulas were made by U.S. astronomer Edward Emerson Barnard (1857–1923; pictured below), who had a sharp eye for spotting elusive celestial objects and astronomical phenomena. However, his important astronomical discoveries were made without much benefit from formal schooling. Barnard's father died before his birth, and at the early age of nine, Barnard had to go to work in a portrait studio to support his family. However, the job gave Barnard extensive experience with the emerging art of photography, a skill he later exploited scientifically by taking photographs of celestial objects, effectively pioneering the art of celestial photography.

From the age of 26, Barnard spent four years at Vanderbilt University in Tennessee. Even before arriving at the school, he had discovered several comets, and he added more to his list of discoveries while he was a student. He left the university in 1887—without having earned a degree—to join the staff of the then-new Lick Observatory in California. Barnard took full advantage of Lick Observatory's large telescope to take photographs of the Milky Way galaxy (see MILKY WAY), which revealed many new nebulas. He made many discoveries, but his most important one, in 1892, was of Jupiter's fifth moon (see JUPITER).

In 1895 Barnard moved from the Lick Observatory to take a professorship at the University of Chicago's Yerkes Observatory in Wisconsin, where he deduced that the darkness in dark nebulas represents clouds that are blocking light from other celestial objects.

Barnard died in 1923. Among the memorials to him are Barnard's Loop, the name given to a spherical mass of hydrogen that surrounds the Orion Nebula. This structure formed when energy from hot, young stars pushed outward on thin interstellar gases surrounding the stars, forming an interior void and an external shell of hydrogen gas.

DISCOVERERS

which looks like an elliptical ring in space. Many planetary nebulas have a detectable star at their center, and astronomers believe that the rings are actually shells of gaseous material that the star has blown into the surroundings. When viewed from Earth, perspective makes the three-dimensional, translucent shell resemble a flat ring.

The planetary nebula is produced toward the end of the life span of any star similar in size to the Sun. Gradually, the star uses up much of the hydrogen in its core, and it balloons into a type of star known as a red giant (see RED GIANTS). Eventually, the star begins to fuse helium atoms instead of hydrogen atoms in its core. This unstable process causes the star to pulsate wildly, until a particularly violent pulsation ejects the star's coolest outer layers into space. The cool material becomes a planetary nebula that continues to expand into space. The surviving part of the star becomes a white dwarf (see WHITE DWARFS), and its radiation causes the surrounding planetary nebula to glow until the white dwarf cools so much that the radiation emitted is not of a high enough energy level to allow the planetary nebula to continue to glow.

Nebulas can be illuminated by other than visible light, in particular by X rays (see X RAYS). The Crab Nebula (also known as M1) in the constellation Taurus is one such emission nebula. The Crab Nebula—named because the tendrils extending from its cloud resemble a crab's legs—is the aftermath of the explosion of a supernova (see NOVAS AND SUPERNOVAS) in the year 1054. The nebula consists of gas that the explosion spewed into space. The star at its heart collapsed to form a pulsar, which is a neutron star that spins 30 times every second (see PULSARS).

Reflection nebulas

A reflection nebula reflects light from stars toward Earth, so the reflection nebula appears bright. Dust inside a reflection nebula is illuminated by light from a nearby star or even by the collective light of the galaxy as a whole. Reflection nebulas often have a bluish color, because blue light is more efficiently scattered by dust particles than longer-wavelength light. A prime example of a reflection nebula is the Pleiades cluster, M45. Light from the Pleiades, an open cluster of 3,000 or more stars, is scattered by dust grains in a nebula, and the nebula glows with a characteristic blue light.

V. KIERNAN

See also: BLACK HOLES; CONSTELLATIONS; DARK MATTER; GALAXIES; HYDROGEN; IONS AND RADICALS; JUPITER; LASERS AND MASERS; MILKY WAY; NOVAS AND SUPERNOVAS; PULSARS; QUASARS; RED GIANTS; ULTRAVIOLET RADIATION; WHITE DWARFS; X RAYS.

Further reading:

Kwok, S. 2001. *Cosmic Butterflies: The Colorful Mysteries of Planetary Nebulae.* New York: Cambridge University Press.

NEPTUNE

Neptune is a cold, blue-green planet with an atmosphere rich in hydrogen, helium, water, and methane

Neptune is named for the Roman god of the seas. It is usually the eighth planet from the Sun but, at times, becomes the most distant when Pluto takes an occasional 20-year-long swing inside Neptune's orbit.

Neptune is one of four giant, or Jovian, planets, the others are Jupiter, Saturn, and Uranus. Sometimes thought of as a twin of Uranus, Neptune is only superficially similar. Although both planets have hydrogen as their major atmospheric component and appear blue-green through a telescope, Neptune is more massive, has a higher density, and, unlike Uranus, has an internal heat source (see URANUS). Neptune radiates more than twice as much energy as it receives from the Sun, as do Jupiter and Saturn.

Neptune has not been tracked over a complete orbit because its orbital period is about 165 years, and it was detected only in 1846. However, Neptune was first sighted much earlier. Italian astronomer Galileo Galilei (1564–1642) made some observations of Neptune soon after the invention of the telescope.

Neptune's average density is the highest of the Jovian planets, suggesting a different composition and internal structure from the other planets. Speculative models of the structure of Neptune suggest a central, perhaps molten, rocky core surrounded by ices of water, methane, and liquid ammonia. At higher altitudes, a gaseous envelope of hydrogen and helium also reveals the characteristic blue-green of methane.

A first real look at Neptune

Because of Neptune's extreme distance from Earth, detailed planetary information, such as atmospheric structure, had to await a spacecraft visit. NASA's *Voyager 2* flight in 1989 (see NASA) revealed a turquoise planet encircled by narrow rings and streaked by white clouds. Several atmospheric features persisted. A huge storm, the Great Dark Spot, spun counterclockwise in the atmosphere as it swept across Neptune at nearly 750 miles per hour (1,200 km/h). A small, white cloud, nicknamed Scooter, circled the planet once every 16 hours.

The atmosphere of Neptune has winds that blow at 1,250 miles per hour (2,000 km/h), which are among the fastest in the solar system (see WIND). Most of these winds blow westward, opposite to the direction of Neptune's rotation.

Unlike those of the other Jovian planets, Neptune's clouds generally move around the planet more slowly than the planet rotates. They power an atmosphere that may change as rapidly as Earth's. For example, *Voyager 2* saw shadows of higher-altitude clouds on lower cloud decks. Both the Great Dark Spot and Scooter had diminished, if not disappeared by, June 1994, when the Hubble Space Telescope (see SPACE EXPLORATION; TELESCOPES) took a good look at Neptune. Five months later, Hubble Space Telescope images showed a new dark spot apparently forming in the northern hemisphere. High-altitude clouds around the edge of this new spot were perhaps cooled clouds of methane ice crystals. Some astronomers suggest that the dark spot itself may be a window to a cloud deck lower in the atmosphere.

This view of Neptune was taken during the orbit of* Voyager 2 *in August, 1989. The Great Dark Spot, on the left in this picture, is approximately the same size as Earth and is surrounded by white, wispy clouds of methane.

NEPTUNE

- Distance from the Sun: 2.8×10^9 miles (450×10^9 km)
- Orbital period: 164.8 Earth years
- Rotation period (day): 16 hours 7 minutes
- Eccentricity: 0.0082
- Equatorial diameter: 30,780 miles (49,530 km)
- Mass: 2.27×10^{26} pounds (1.03×10^{26} kg)
- Average density: 101.93 pounds/cubic feet (1.64 g/cm^3)

CORE FACTS

- The planet Neptune was discovered in 1846. It had been previously observed but was thought to be a star.
- Neptune is the eighth farthest planet from the Sun, but Pluto occasionally swings inside its orbit.
- Neptune is the densest of the four Jovian planets.
- Neptune has an atmosphere of hydrogen, helium, water vapor, and methane.

CONNECTIONS

- In ascending order of distance from the **SUN**, the four Jovian planets are **JUPITER**, **SATURN**, **URANUS**, and Neptune.
- Longer wavelengths of **LIGHT** are absorbed by Neptune's **ATMOSPHERE**, and thus the planet appears blue-green.

THE DISCOVERY OF NEPTUNE

Uranus, which was discovered in 1781, appeared to speed up and slow down unpredictably, and this behavior puzzled astronomers. Observations of Uranus showed it ahead of its predicted position, and then, suddenly in 1829, it was behind where it should have been. Two young scientists worked out the answer to the riddle of Uranus's strange motion independently: French mathematician Urbain-Jean-Joseph Le Verrier (1811–1877) and British astronomer John Couch Adams (1819–1892). Both predicted the presence of an as yet unseen planet whose gravity was affecting the path of Uranus.

Unfortunately for Adams, George Biddell Airy (1801–1892), the astronomer royal at the time, rebuffed his hypothesis of an undiscovered planet. British astronomers searched briefly for Neptune in July and August 1846. They actually saw Neptune but failed to realize they had discovered the planet. Meantime, Le Verrier wrote to German astronomer Johann Gottfried Galle (1812–1910) at the Berlin Observatory, telling him where to look for the mystery planet. Galle received the letter on September 23, 1846, and that night he and student Louis d'Arrest found Neptune. It was exactly where the calculations of both Le Verrier and Adams said it should be. However, because Airy had effectively buried Adams's studies, it was some years before Adams received equal credit for the discovery of Neptune.

HISTORY OF SCIENCE

Neptune's orbit is almost circular. The axis of rotation is considerably inclined to the orbital plane, so seasons do occur (see SEASONS). In fact, Neptune's brightness (the light it reflects from the Sun) appears to change over periods of months or years.

Neptune has five rings, mostly so thin that they are difficult to see from Earthbound telescopes. One of the rings is unusual because it is made up of unconnected arcs of material. Similarly to the other Jovian planets, it has a magnetic field and a magnetosphere.

The moons of Neptune

Only two of Neptune's eight known moons had been discovered before *Voyager 2*'s arrival in 1989 (see SATELLITES, NATURAL). Most are small and irregular in shape, suggesting that they are bodies that have been captured from space rather than true moons.

Triton, a moon that has been known since shortly after the discovery of Neptune, circles close above the planet. As its orbit is retrograde (opposite to the planet's rotational direction), astronomers suggest that Triton was not formed with Neptune but may instead have been captured into Neptunian orbit at a later date. Some have suggested that Triton is a lot like Pluto (see PLUTO). Triton is spiraling toward Neptune, pulled by tidal interactions between the satellite and the planet. Billions of years from now, astronomers predict that Triton orbit will fall into Neptune's atmosphere and be destroyed.

Triton appears to be similar to Ganymede and Callisto, two of Jupiters moons, suggesting that it is rocky rather than icy. The surface temperatures of less than −390°F (−235°C) account for the ice caps that cover some of Triton's surface. Triton and Titan, Saturn's giant satellite (see SATURN), are the only satellites known to have significant atmospheres.

J. DENNETT

See also: ASTRONOMY; COSMOLOGY; NASA; PLUTO; SATELLITES, NATURAL; SEASONS; SOLAR SYSTEM; SPACE EXPLORATION; TELESCOPES; WIND.

Further reading:

Miner, Ellis D., and Randii R. Wessen. 2002. *Neptune: The Planet, Rings, and Satellites.* New York: Springer.

A Voyager 2 image of Neptune's largest moon, Triton. The surface of the moon, dubbed the cantaloupe terrain, is covered with small dimples with upraised rims and shallow central depressions.

NEUTRINOS

Neutrinos are subatomic particles with no charge and a near-zero mass

According to Frederick Reines, one of the physicists who first experimentally confirmed the existence of neutrinos, the neutrino is "the most tiny quantity of reality ever imagined by a human being." Uncharged and almost massless, these extraordinary small particles barely interact with matter at all.

Fundamental particle

Neutrinos belong to the group of subatomic particles called leptons (see LEPTONS). These matter particles, which also include electrons and muons, are thought to be fundamental—that is, they are not made up of any other particles (unlike protons, for example, which are made up of smaller particles called quarks; see QUARKS).

Three types of neutrinos are known: the electron neutrino, the muon neutrino, and the tau neutrino. Thus, there is a neutrino for each of the charged leptons. The relationship of each neutrino to the matching charged lepton is that they occur together in subatomic particle reactions. For example, in radioactive beta decay, a neutron changes to a proton releasing an electron and an electron neutrino, and when the unstable muon and tau particles decay, a matching muon or tau neutrino is produced (see RADIO WAVES AND RADAR).

In terms of their mass, lifetime, and interactions, all three neutrinos are similar. They are stable and for a long time were thought to be massless, although recent evidence suggests they do have some mass. All neutrinos take part only in gravitational and the so-called weak interactions (see FORCES; GRAVITY).

Elusive matter

Although they exist in vast numbers in the universe—several hundred million in every cubic yard of space, or one million times the number of photons (particles of light)—their interaction with matter is so slight that detecting neutrinos at all has taxed the ingenuity of experimental physicists. Neutrinos can pass through virtually anything. For example, a single neutrino would have only a 50 percent chance of being stopped by 100 million miles (160 million km) of solid steel. The huge majority pass straight through Earth and even through the Sun.

Neutrinos were first proposed during the 1930s to solve a puzzle. Researchers noticed that after radioactive beta decay, the sum of the masses of the resulting particles (a proton and an electron) was lower than the mass of the original particle (a neutron)—a violation of the law of conservation of matter (see THERMODYNAMICS). Austrian physicist Wolfgang Pauli (1900–1958) proposed in 1931 that another particle must be produced to account for the missing mass. This idea was accepted by physicists, and in 1934 Italian physicist Enrico Fermi named this particle the neutrino, meaning "little neutral one."

An artwork showing some of the optical modules of a neutrino telescope planned by French researchers. The instrument is being built in the Mediterranean Sea and will consist of 10 vertical strings, each equipped with 100 detectors. The detectors look for radiation produced by muons in the water. A muon produces radiation after it interacts with a neutrino.

The electron neutrinos were thus the first neutrinos known, but experimental verification of their existence had to wait until 1956. In that year Frederick Reines (1918–1998) and Clyde Cowan (1919–1974) used the United State's newly developed nuclear reactors as a good source of these elusive particles. They built a detector close to a nuclear reactor at Savannah River in South Carolina, siting it underground to screen out the other particles from the reactor. Reines and Cowan managed to detect several reactions that they showed involved antiparticles to what are now called electron neutrinos.

Since then, physicists have started to use the remarkable properties of neutrinos to track events in particle accelerators and in the cosmos. In 1987 neutrinos from a supernova (caused when a large star explodes; see NOVAS AND SUPERNOVAS) were detected on Earth. This discovery occurred before astronomers saw the supernova through telescopes. In a period of 10 seconds, a total of 19 neutrinos were detected—a remarkably rapid rate that meant that 65 billion neutrinos hit each square inch (10 billion/cm^2) of the Earth's surface. Bearing in mind that the supernova occurred 170,000 years ago and that the neutrinos had been spreading out in all directions from the star for this time, the number of neutrinos released was literally astronomical.

S. WATT

See also: ELECTRONS AND POSITRONS; FORCES; GRAVITY; LEPTONS; NEUTRONS; NUCLEAR PHYSICS; PARTICLE PHYSICS; QUARKS; RADIOACTIVITY; RADIO WAVES AND RADAR; THERMODYNAMICS.

Further reading:
Gribben, John R. 2000. *Case of the Missing Neutrino and Other Curious Phenomena of the Universe*. New York: Berkley Publishing Group.

CONNECTIONS

- Because neutrinos are so light and numerous, they are thought to be a major component of invisible **DARK MATTER**.

- **NOVAS AND SUPERNOVAS** are sources of huge quantities of neutrinos.

NEUTRONS

Neutrons are neutral particles that are confined within the nucleus but can be freed in certain nuclear reactions

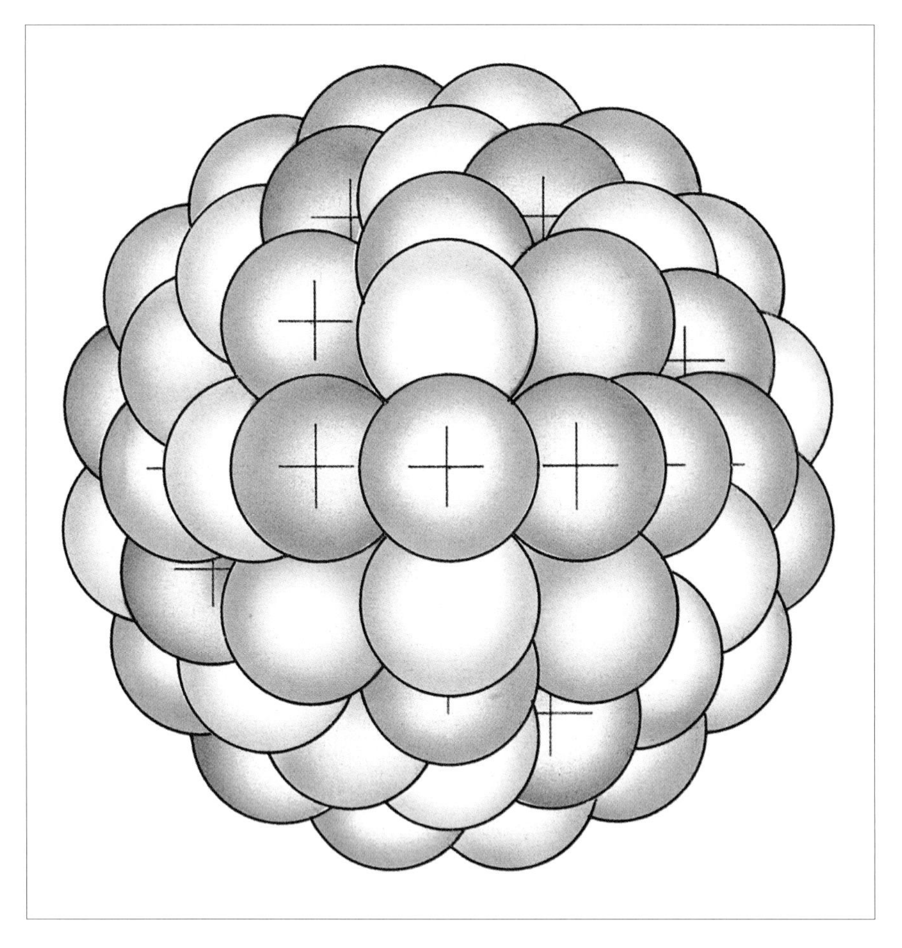

A nucleus consists of tightly packed positively charged protons (red) and uncharged (electrically neutral) neutrons (blue).

An atom can be pictured as a tiny solar system, with neutrons and protons forming its nucleus and a cloud of orbital electrons surrounding it. A nucleus would not hold together without the neutrons. The protons inside the nucleus are positively charged. They have a short-range attraction for one another, but the repulsion between the protons stemming from their similar electric charges overcomes this attraction. Neutrons have a strong short-range attractive force for both protons and other neutrons, so they hold all the particles together.

Every chemical element has a specific number of protons in the nucleus, which is equal to the number of electrons surrounding it in the neutral atom, but the number of neutrons is variable. For example, iron atoms always contain 26 protons, but they can have 28, 30, 31, or 32 neutrons, forming four stable isotopes (see ISOTOPES). By bombardment with free neutrons, other isotopes not found in nature can usually be produced. However, these are unstable and disintegrate over time.

In special circumstances, neutrons can be liberated from the nucleus. In 1932 British physicist James Chadwick (1891–1974) used alpha particles emitted from a radioactive source to bombard the element boron (see RADIOACTIVITY). He observed that occasionally some radiation was produced that could penetrate more than 0.4 inches (1.2 cm) of lead. From past experience, he knew that electrically charged particles would be stopped by a thickness of less than a millimeter. He suggested that the unusually penetrating radiation was made up of neutral particles, or neutrons.

In an earlier experiment, Ernest Rutherford (1871–1937) at England's University of Manchester had shown that energetic particles from a radioactive source can enter the nucleus of an atom and expel a proton from it. From this observation, Chadwick reasoned that a similar nuclear reaction might knock a neutron out of a nucleus.

Neutral particles do not leave a track in a cloud chamber, nor can they be detected with a Geiger counter. What Chadwick did was to place a layer of paraffin behind the lead. Neutrons coming through the lead would collide with hydrogen nuclei (protons) in the paraffin.

As in a pool ball collision, the protons recoiled, left the paraffin, and produced counts in a detector. In 1935 Chadwick received the Nobel Prize in physics for his discovery of neutrons.

Nuclear reactions

Enrico Fermi (1901–1954), a young Italian-born scientist, followed up on Chadwick's experiments. Using a strong source of radium to bombard the element beryllium, he was able to produce more than a million neutrons per second. When various materials were exposed to these neutrons, the materials were found to have become radioactive.

For example, when bombarded with neutrons, the stable element cobalt becomes radioactive cobalt-60, which is now widely used in radiation therapy for cancer patients. Radioactive iron, produced by neutron bombardment of ordinary iron, has been used to measure the wear of steel parts in automobile engines.

One isotope of uranium, the heaviest element in nature, was found to have a unique nuclear reaction when bombarded by neutrons. The uranium nucleus breaks into two pieces, releasing a large amount of energy. In this process, which is called fission, several neutrons are also emitted (see FISSION). These

CONNECTIONS

- **NEUTRON STARS**, with their small size and enormously strong **GRAVITY**, have provided a good explanation for **PULSARS**.
- The **QUARK** model has been used by scientists to explain the magnetic properties and radioactive decay of neutrons.

CORE FACTS

- Neutrons are electrically neutral particles.
- Neutrons interact with protons and with each other to provide binding energy for the nucleus.
- Radioactive isotopes for medicine and industry can be created by bombarding ordinary elements with neutrons.
- When the uranium-235 isotope is bombarded with neutrons, its nucleus can split apart with a large release of energy and the emission of several neutrons.
- Neutrons consist of three quarks.

secondary neutrons can enter other uranium nuclei, causing more fissions and releasing more neutrons. Such a chain reaction, creating an explosion, was the basis for the atomic bomb developed during World War II (1939–1945). There are two isotopes of uranium, and it is the scarcest, U^{235}, that undergoes this fission reaction (see NUCLEAR PHYSICS). The most common isotope is U^{238}. It is an enormous technological effort to separate these isotopes.

Certain materials, including the element cadmium, are good neutron absorbers. A chain reaction can be kept under control by using cadmium rods to hold the number of neutrons constant. In this way, a steady-state condition can be maintained in the operation of nuclear power plants.

Although neutrons are electrically neutral, they have some peculiar properties. For example, a beam of neutrons is deflected when passing through a sheet of magnetized iron. That is, they react like tiny magnets (see MAGNETISM). Also, free neutrons undergo radioactive decay. Such observations suggest that neutrons are not truly fundamental particles but have some kind of substructure.

In 1964 U.S. physicist Murray Gell-Mann (b. 1929) proposed the so-called quark model for nuclear particles (see MESONS; QUARKS). According to this model, a neutron would consist of three quarks, one "up" quark having a charge of $+\frac{2}{3}$ and two "down" quarks, each with a charge of $-\frac{1}{3}$, the result being a net charge of zero. The quark model has been used to explain the magnetic properties and radioactive decay of neutrons (see RADIOACTIVITY). Experimental evidence for quarks inside the neutron was first obtained in the 1970s from electron-scattering experiments at the University of Stanford accelerator, although no isolated quark has ever been observed because the forces binding quarks are so strong (see PARTICLE ACCELERATORS).

Neutron stars

The concept of a neutron star was proposed in 1939 by U.S. physicist J. Robert Oppenheimer (1904–1967), who later became head of the atomic bomb project at Los Alamos Laboratory, New Mexico. Oppenheimer proposed that some dying stars can undergo a cataclysmic event called gravitational collapse. Intense gravitational pressure would force orbital electrons inward toward the nucleus, where they could combine with protons to form neutrons. A star somewhat larger than the Sun would become a concentrated cluster of neutrons only a few miles in diameter after gravitational collapse.

Neutron stars remained only a theoretical possibility until the 1960s, when astronomers discovered the first pulsars, which emit regular pulses of radiation less than a second apart (see PULSARS). Such regularity was attributed to stellar objects rotating very rapidly. A normal star could not spin so fast, because centrifugal force would tear it apart. However, pulsars may be neutron stars. With its small size, the neutron star could easily rotate very rapidly.

H. GRAETZER

JAMES CHADWICK

James Chadwick (1891–1974) was a graduate student in physics under Ernest Rutherford (1871–1937) in 1911, when the nuclear atom model was first established. He was carrying out research in Berlin with German physicist Hans Geiger (1882–1945), who invented the Geiger radiation counter, when World War I (1914–1918) broke out, and he was interned for four years. Even so, he managed to continue with some research projects. After the war, Chadwick rejoined Rutherford and continued to investigate nuclear reactions. His discovery of a new particle, the neutron, earned him the Nobel Prize in 1935. In 1939 he was asked by the British government to evaluate the possibility of a chain reaction based on uranium fission, and from 1943 to 1945, he coordinated British contributions to the atomic bomb project. He was knighted in 1945.

DISCOVERERS

See also: ATOMS; ELECTRONS AND POSITRONS; FISSION; ISOTOPES; MAGNETISM; MESONS; NEUTRINOS; NUCLEAR PHYSICS; NUCLEOSYNTHESIS; PARTICLE ACCELERATORS; PARTICLE PHYSICS; PROTONS; PULSARS; QUARKS; RADIOACTIVITY.

Further reading:

Lilley, J. S. 2001. *Nuclear Physics: Principles and Applications.* Hoboken N.J.: John Wiley & Sons.
Seiden, Abraham. 2004. *Particle Physics: A Comprehensive Introduction.* San Francisco, Calif.: Addison Wesley.

NEUTRON STARS

Neutron stars are small, incredibly dense stars formed in the final stage of a massive star's life

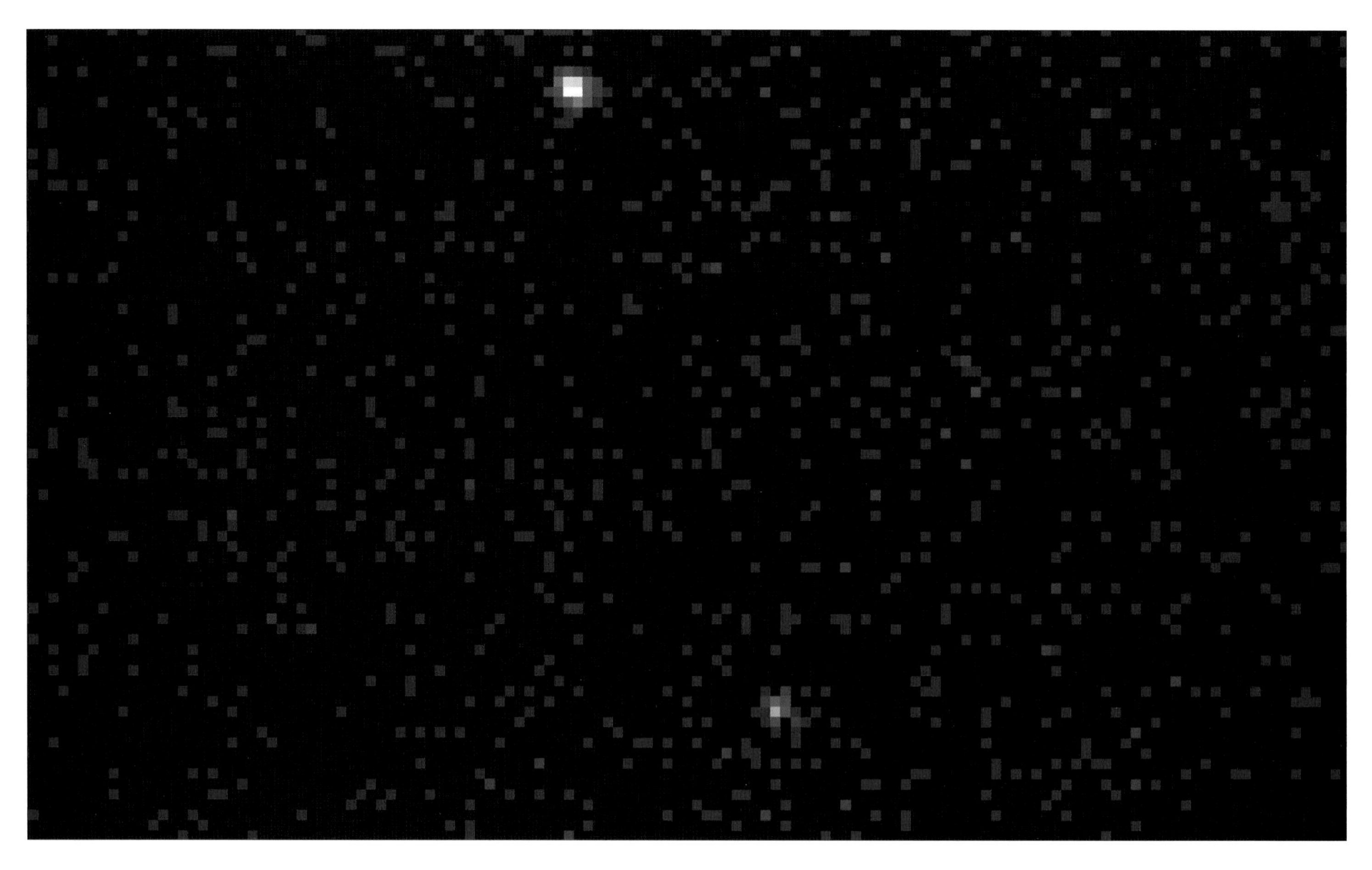

Neutron star KS 1731-260 (at the top of the picture) is glowing at 5.4 million°F (3 million°C). This is a low temperature for a neutron star and may have only recently heated up. Astronomers think there may be many more colder neutron stars in the sky that are too dim to see.

Neutron stars are very unusual stars unlike any other type of star (see STARS). To start with, they are tiny, measuring only about 10 to 12.5 miles (15 to 20 km) across. Yet, they are made of more material than the Sun. A neutron star is typically made of about 1.4 to about 3.2 solar masses worth of material, which is packed very tightly into a small volume (see DENSITY). The material is so dense that a piece the size of a grain of sand would weigh 4 million tons (3.6 metric tons) on Earth. Neutron stars are simply the densest stars known. They stand out from the crowd in other ways, too. Unlike other stars, which are spheres of gas without a confining edge, the neutron star has a solid surface. More intriguing still is the material they are made of.

The name of the star gives a clue to its makeup. The neutron star is primarily made of neutrons, which are present in the nucleus of all atoms except hydrogen (see ATOMS; NEUTRONS). The star's material was not always in this form. The original star has undergone gravitational collapse, and its material has been compressed into neutrons. A neutron is formed when a proton and an electron combine under extreme gravitational pressure. The resulting star has strong gravity and an intense magnetic field and spins rapidly (see GRAVITY; MAGNETISM). Astronomers know of about 2,000 such neutron stars, and this number is rising rapidly.

Anatomy of a neutron star

British astronomer Jocelyn Bell Burnell (b. 1943) has compared the neutron star to a raw egg. They both have solid, strong shells that enclose gooey liquid. From the outside in, the neutron star has a few-hundred-foot-thick crust of polymers made from iron atoms (see POLYMERS). Below is a neutron-rich area, and the deeper one goes, the richer the material becomes. The deepest layer is a densely packed sea of neutrons. It is a superfluid that flows without any friction (see FRICTION; LIQUIDS). The makeup of the central core is unknown. Some cores might be solid, while others might be liquid. It is known, however, that the core includes some protons and electrons as well as neutrons. Without them the neutron star could not support its magnetic field.

CONNECTIONS

- **PULSARS** are spinning neutron stars.
- Neutron stars are detected by the **RADIO WAVES** or **X RAYS** they emit.

CORE FACTS

- Everything about neutron stars is extreme. They are the smallest, densest and fastest-spinning stars, and unlike other stars, they have a solid surface.
- Neutron stars are made primarily from neutron particles, which are densely packed in a city-sized sphere.
- Neutron stars are the final stage in a massive star's life. These stellar corpses are found in supernova remnants and in binary stars.

When a neutron star is created, it shrinks in size, and its surface area is much reduced. The original star's magnetic field, which was once spread across a large surface area, is now concentrated on a much smaller one. Consequently, the strength of the field increases. The magnetic field of a neutron star is typically one trillion gauss compared with that of an average star, which is typically in the range of one to a few thousand gauss. (A field of 1,000 gauss produces a force of 1 newton per ampere meter.) The strength of the field is important because it enables the neutron star to emit pulses of radiation.

Spinning neutron stars

All stars rotate, and neutron stars are no exception. When a star shrinks, its spin rate increases, and since neutron stars are so small, they spin incredibly quickly. Take the Sun, for example; if it were to shrink to neutron star size, its rate would increase from its present rate of one spin each month to about a thousand spins each second. The density of a neutron star stops it from flying apart as it spins. The gravitational pull of the star is so strong that anything trying to get away from its surface would need to travel (have an escape velocity) at about half the speed of light.

The effect of gravity on anyone standing on a neutron star, were such a thing possible, is equally extreme. The gravitational pull increases in strength the closer a body is to the surface. The pull on the feet is stronger than that on the legs, which is stronger than that on the torso, and so on. The body is pulled apart as a result. Anyone standing on a neutron star would not have time to look around, but if a person could stand on a neutron star, the view would be surprising. Because light on the star's surface is bent by the gravity, an observer would be able to see tens of degrees over the horizon.

The intensity of a neutron star's magnetic field also produces extraordinary effects. The star's rapid rotation and its strong magnetic field work together as an electrical generator. The result is high-energy particles, which produce beams of radiation. The radiation energy is channeled away from the star by the magnetic field and then sweeps across space as the star rotates. Two narrow beams of radiation stream out of the star's north and south magnetic poles. A line connecting the two poles, the star's magnetic axis, is tilted from the star's rotation axis. So, as the star spins, the beams of radiation sweep around the sky.

Imagine a spinning ice skater whose body stays perfectly vertical as she spins, her head pointing to the same spot in the sky above. Now imagine she holds a torch aloft in an outstretched arm and points it to a patch of sky not directly above her head. As she spins, keeping her arm in the same relative position to her body, the torchlight sweeps across the sky in much the same way that a beam of radiation does as it emits from one of the magnetic poles of a neutron star. Anyone in the path of the torchlight would see a flash of light each time the skater spun round. The

JOCELYN BELL BURNELL

In the late 1960s Jocelyn Bell (as she was then named) was a student at Cambridge University. She was part of a small team that over two years connected 120 miles (196 km) of cable and wire to more than 1,000 posts over a 4.5 acre (1.8 ha) site. The result was an array of radio antennae, which Bell then used to pick up radio emissions from space. She would set the array to scan the sky, strip by strip, completing one scan every four days. On behalf of the team, she searched for quasars, which, as compact objects, scintillated (twinkled) in the radio sky (see GALAXIES; QUASARS).

Bell noticed, on more than one occasion, a "funny, scruffy signal' in among her results. The signal was radio pulses coming every 1.34 seconds from a point within Earth's galaxy, located in the constellation of Vulpecula (see CONSTELLATION). The discovery of this source, now called CP 1919, was made on August 6, 1967. It was observed again in November. The rapidity and regularity of the pulses suggested they may be of alien origin. With this idea in mind, the source was at first designated LGM 1 (for Little Green Men). However, when over the next two months three more sources were found, it was thought to be highly unlikely that intelligent beings in different locations were sending out simultaneous signals. The LGM designations were discarded and astronomers looked for other explanations for the signals. In February 1968 the discovery of the first pulsating source was announced. A subsequent radio interview with the science correspondent of the U.K. newspaper the *Daily Telegraph* produced the name for these new objects. On hearing that Bell (pictured below) and her supervisor Tony Hewish had no name, he suggested *pulsar*, short for "pulsating radio star." Jocelyn Bell Burnell has worked as an astronomer ever since she discovered pulsars. She is now Professor Jocelyn Bell Burnell, President of the Royal Astronomical Society, and Dean of Science at Bath University, England.

DISCOVERERS

BIRTH OF A NEUTRON STAR

Neutron stars are the collapsed cores of massive supergiant stars (see STARS). Such stars, which have more than about eight times the Sun's mass, can end their lives in a supernova explosion. Throughout its lifetime the supergiant has converted the material in its core to heavier and heavier elements (see NUCLEOSYNTHESIS). The final element produced is iron. When the core is made of about 1.4 solar masses worth of iron (the Chandrasekhar limit; see CHANDRASEKHAR, SUBRAHMANYAN), the star becomes unstable, and its core collapses. This collapse is followed almost immediately by the rest of the star exploding.

The core contracts, its protons and electrons merge to form neutrons, and a neutron star is formed. A neutron star is not produced, however, if the mass of the contracting material is much greater than about three solar masses. If it is, contraction continues past the neutron star stage, and a black hole is formed (see BLACK HOLES). In this way neutron stars have an upper limit to their mass.

A neutron star can also form from a white dwarf, which increases its mass by gathering material from another star (see BINARY STARS; WHITE DWARFS). The mass of the white dwarf rises until it reaches the Chandrasekhar limit; the star then collapses, and a neutron star results.

A CLOSER LOOK

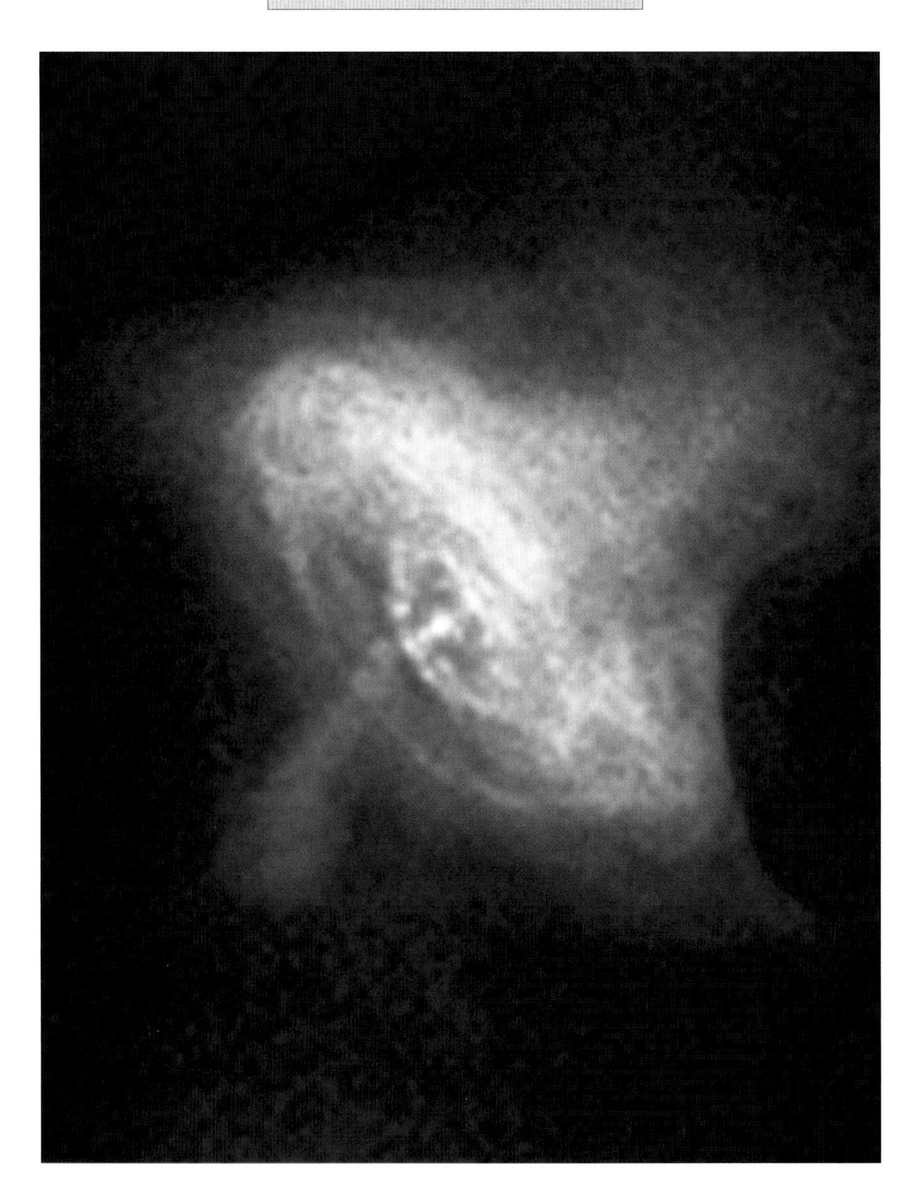

The Crab Nebula in the constellation of Taurus is the remains of a supernova explosion that occurred 1,000 years ago. At its heart is a neutron star.

light would appear to be flicking on and off alternately as the skater spun, even though it would be shining all the time. Similarly, if Earth lies in the path of a beam of radiation emanating from a neutron star, observers detect the radiation, which pulses on and off as the star turns.

A neutron star detected in this manner is called a pulsar (see PULSARS). The rate of the pulses detected tells astronomers that such neutron stars spin typically in the range of hundreds of times a second to once every few seconds. The fastest of all are the millisecond pulsars, which make as many as a thousand revolutions per second. It is believed there are many more pulsars than the ones currently detected. Many remain elusive because Earth does not lie in the line of sight of their radiation beams (see DARK MATTER). It is estimated that astronomers are aware of only about one in every five.

The rotations of pulsars have slowed down steadily over thousands of years. A pulsar can, however, suddenly and for a brief period speed up. This short-lived acceleration is known as a glitch. The crust of the neutron star slows, but the layer underneath continues to spin at its original rate. The interaction between the two layers creates a jolt resulting in a sudden and brief spurt of speed from the crust, after which the pulsar returns to its usual rate of slowing down.

Prediction and discovery

The idea of the neutron was established in the early 1930s. The atomic particle, the neutron, was discovered by English physicist James Chadwick (1891–1974) in 1932. Earlier in the same year Soviet astronomer Lev Landau (1908–1968) had proposed the idea of a collapsed star with the density of a nucleus, although at the time it was not called a neutron star. In 1933 Bulgarian-born Fritz Zwicky (1898–1974) and German Walter Baade (1893–1960) invented the concept of the supernova (see NOVAS AND SUPERNOVAS). They suggested that a supernova was the transition from an ordinary star to a neutron star. The theory of neutron stars followed in 1939 when U.S. physicist Robert Oppenheimer (1904–1967) and Canadian George Volkoff (1914–2000) published their work, which showed that the mass of a neutron star cannot be more than 3.2 suns (see BLACK HOLES).

The neutron star was regarded during the 20th century as something very peculiar, and their existence was not pursued. Things changed after 1968, however, when the discovery of pulsating radio sources, pulsars, was announced (see RADIO WAVES AND RADAR; TELESCOPES).

Astronomers were not sure at first what they had found. It was suggested they were rapidly rotating white dwarf stars (see WHITE DWARFS). In the 1960s white dwarfs were thought to be the only type of stellar corpse (see ASTROPHYSICS). However, it was soon realized they could not be. White dwarfs were too large to spin at the speed suggested by the newly discovered pulsars; the dwarfs would rip themselves

apart. Pulsars must be smaller bodies, albeit ones containing huge amounts of energy.

The idea of rapidly rotating neutron stars emitting beams of radiation was considered and accepted. Once a pulsar was discovered in the center of the Crab Nebula, in November 1968, the nature of pulsars became understood. The pulsar and the nebula were the remains of a supernova that exploded in 1054 C.E. (see NOVAS AND SUPERNOVAS). Other neutron stars have since been discovered at the heart of supernova remnants.

Neutron stars are usually detected by their radio pulses, as pulsars. However, a small number have been discovered as neutron stars, most notably by X-ray satellite observation since the 1990s. However, Sco X-1, a neutron star in a binary system (one involving two stars) was discovered way back in 1962, before the first pulsar discovery (see BINARY STARS). Yet it was not until the early 1970s that it and a few dozen others were confirmed as neutron stars.

Neutron stars in binary systems have a normal star as their companion. Material discarded by the normal star is captured by the neutron star and forms a disk around it. The material then spirals down onto the smaller star's surface. Consequently, energy is produced, and much of it is radiated away as X rays (see X RAYS). When the energy release is concentrated in a column or spot, it can be detected as X-ray pulses as the neutron star rotates. In some cases, the pulse is steady; in others, there is an explosive burst. Neutron stars in binary systems are some of the strongest X-ray sources in the Milky Way, Earth's galaxy (see MILKY WAY). The Chandra X-ray Observatory satellite has detected hundreds more of these X-ray binaries in other galaxies too.

Astronomers know of six binary systems that are made up of two neutron stars. The sixth, designated PSR J0737-3039, was discovered in December 2003 by astronomers making a routine pulsar search with the Parkes Radio Telescope in Australia. Equipment installed on that telescope in the late 1990s has so far enabled astronomers to discover about half of all pulsating neutrons known. One of the new pair is a pulsar; the second remains unseen, as it does not emit radio pulses in Earth's line of sight. Its presence is inferred by its gravitational effect on its companion. Astronomers predict that the two will collide and merge together in about 85 million years.

C. STOTT

See also: ASTROPHYSICS; ATOMS; BINARY STARS; BLACK HOLES; CONSTELLATIONS; DARK MATTER; DENSITY; FRICTION; GALAXIES; GRAVITY; LIQUIDS; MAGNETISM; MILKY WAY; NEUTRONS; NOVAS AND SUPERNOVAS; NUCLEOSYNTHESIS; POLYMERS; STARS; TELESCOPES; WHITE DWARFS.

Further Reading:

Bell Burnell, J. 2004. Pliers, pulsars, and extreme physics *Astronomy and Geophysics* (February) 7–11.
Fryer, Chris L., ed. 2004. *Stellar Collapse.* Boston: Kluwer Academic Publishers.

An artist's impression of the Chandra X-Ray Observatory. This satellite points away from Earth and scans the heavens as it orbits. It detects X-ray sources, including many previously unknown neutron stars.

NEWTON, ISAAC

Mathematician and physicist Isaac Newton was an important contributor to the development of modern science

As well as working with physics and math, Newton was also interested in chemistry.

> **CONNECTIONS**
>
> - Newton advanced the science of **MATHEMATICS** when he developed **CALCULUS.**
> - The change in position of an object is called its **MOTION**. It is governed by three basic laws put forward by Newton.

Isaac Newton was born prematurely on Christmas Day 1642—according to the Julian calendar, or January 4, 1643, by the current Gregorian calendar—in Woolsthorpe, Lincolnshire, England, and his mother did not expect him to live. His father, the squire of the manor of a small estate, died before Isaac was born.

Newton attended the village school but did not display any sign of his enormous intellect. When he was 12, he went to the King's School in Grantham, Lincolnshire. After four years, his mother took him away to manage their farm. However, Newton was not a good farmer and was soon sent to Trinity College at the University of Cambridge to study to become a preacher. At Cambridge, Newton also studied mathematics and came to the attention of mathematician Isaac Barrow (1630–1677). With Barrow as his tutor, Newton began to show his genius. He won a mathematical scholarship in 1664, and when Newton graduated in 1665, Barrow resigned his chair in favor of his pupil.

Revolutionary achievements

In 1665 the bubonic plague reached Cambridge. The university was closed, and people were sent away to help control the spread of the disease. Newton returned home to Woolsthorpe, where he stayed until the end of March 1666. During this time, he was able to concentrate fully on his studies and to complete work on some of the most revolutionary scientific achievements in history. The advances that Newton made during this period include his three laws of motion; his discovery of calculus, the law of gravity, and several properties of optics; and the invention of the reflecting telescope (see CALCULUS; GRAVITY; MOTION; OPTICS; TELESCOPES).

The Royal Society of London

News of Newton's work in optics reached the prestigious Royal Society of London, and he was elected a member. He was also invited to present his findings in a paper. He presented not only the paper but also his reflecting telescope, which remains one of the society's most prized possessions.

However, English scientist Robert Hooke (1635–1703), a member of the society, claimed that Newton stole the ideas from him. Although an accomplished theorist, Hooke did not have the necessary mathematical and experimental skills to investigate his thoughts. As a result, he often speculated on many ideas without providing any supporting evidence. Hooke's claim began a feud with Newton, which lasted until Hooke's death more than 30 years later. Newton was furious with Hooke and his supporters and nearly withdrew from the Royal Society.

> **CORE FACTS**
>
> - Isaac Newton single-handedly contributed more to the development of science than anybody else in history.
> - Newton produced one of the most important and influential works on physics of all times, *Philosophiae Naturalis Principia Mathematica* (Mathematical Principles of Natural Philosophy) (1687), or *Principia.*
> - He made major advances in astronomy, mathematics, and physics.
> - Newton developed the three laws of motion and the law of gravity and invented calculus and the reflecting telescope.

THE LAWS OF MOTION

The motion of all objects in the universe can be explained using three simple laws, which can always be applied except when the velocity becomes great compared with the speed of light. After Italian astronomer and physicist Galileo Galilei (1564–1642) showed that the ideas of Greek philosopher Aristotle (384–322 B.C.E.) were unsound, no one quite understood how things moved (see GALILEI, GALILEO). Isaac Newton then showed that all motion can be described with three simple laws.

The first law states that a body at rest remains at rest and that a body in motion remains in motion in a straight line and at a constant velocity until acted on by an outside force (see FORCES). The second law states that the force (F) exerted by an object is equal to the mass (m) of the object multiplied by the acceleration (a) experienced when the force is applied: $F = ma$ (see ACCELERATION; FORCES; MASS). The third law states that for every action, there is an equal but opposite reaction.

The first law is also called the principle of inertia, and inertia is defined as the tendency of a mass to remain in uniform motion or at rest and is measured by mass. The more mass an object has, the more inertia it has. For example, it takes more work to start rolling a bowling ball than a tennis ball. Similarly, once moving, a greater force is needed to stop the bowling ball than the tennis ball.

The velocity of an object does not change on its own. To change an object's velocity, some unbalanced force must be applied. The size of this force can be found using Newton's second law. The combination of speed and direction is called velocity. Acceleration is the process of changing an object's velocity, either by changing its speed or by changing its direction of motion (see ACCELERATION). The amount of force needed to cause this acceleration is equal to the amount of acceleration multiplied by the amount of mass. The more mass an object has, the more inertia it has, and the harder it is to accelerate the object.

If a force is being applied, there must be something to apply it. Newton's third law states that if one object exerts a force on a second object, the second object must exert a force on the first object that is equal in magnitude but in the opposite direction. This law explains the interaction between all objects. Normally, one is not consciously aware of this reaction because one object is often much larger than the other, and by Newton's second law, the larger object experiences less acceleration from the same amount of force. This law explains how rockets can work even in a vacuum. A rocket pushes out gases with some force, which then exert an equal and opposite force on the rocket, the result being acceleration.

With the exception of very high velocities, which is covered by German-born U.S. physicist Albert Einstein's (1879–1955) theory of relativity, these three simple rules explain all types of motion (see EINSTEIN, ALBERT; RELATIVITY). Scientists applying these laws have been able to improve substantially their understanding of the universe.

A CLOSER LOOK

Although persuaded to remain a member, Newton refused to submit any more of his findings until after Hooke's death.

Newton's *Principia*

Newton isolated himself in Cambridge until 1684, when English astronomer Edmond Halley (1656–1742) approached him with a question concerning gravity. Much to his surprise, Halley found that Newton had discovered the answer nearly 20 years earlier during the plague years. Halley encouraged Newton to put this and other research into a book and even offered to pay for its publication. Newton accepted the offer and began work on *Philosophiae Naturalis Principia Mathematica* (Mathematical Principles of Natural Philosophy, 1687), generally called *Principia*, perhaps the greatest scientific book of all time.

Newton began writing *Principia* vigorously, often going without food and sleep. With this exceptional effort, he finished the project in just 15 months. Published in 1687, the work is a monument to genius. Nearly all the advances in science and technology since the 17th century can be traced to *Principia*. In this work, Newton transcended all the advances achieved by great scientific minds, such as Aristotle and Galileo, and produced a scheme of the universe that was more consistent, elegant, and intuitive than any proposed before (see GALILEI, GALILEO). Newton stated explicit principles of scientific methods, which applied universally to all branches of science. However, the great effort to produce the

A replica of the reflecting telescope invented by Newton in 1671. The telescope collects light with a large mirror and reflects the light into an eyepiece.

PHYSICS AND CALCULUS

Calculus, the mathematics of change and motion, provides a way in which to express the laws of physics in precise mathematical terms (see CALCULUS; MATHEMATICS). In 1665 Newton wanted to investigate some phenomena of nature and found that the mathematics available to him were inadequate. Proving that he was one of the most capable mathematicians in history, Newton set about inventing the required mathematics. What he called fluxions is now called calculus, a Latin word meaning "a stone or pebble used in reckoning."

Calculus was a major advancement in mathematics. It is the branch of mathematics that deals with change and motion by treating a continuously changing function as if it consisted of many small incremental changes. As the size of the incremental changes becomes infinitely small, the value of the function approaches the true value.

There are two forms of calculus, called differential calculus and integral calculus. Differential calculus finds the rate at which a known variable is changing. For example, if the mathematical expression for the velocity of an object is known, differential calculus can be used to find the object's acceleration (the rate of change of velocity) at any given moment. Integral calculus works in the opposite direction; it finds the function when the rate of change is known. For example, if one knows the acceleration of an object, integral calculus provides the expression for the velocity.

Calculus provides a way of easily expressing physical laws with great precision. German astronomer Johannes Kepler (1571–1630), for example, spent more than 20 years figuring out his three laws of planetary motion (see COSMOLOGY). However, by using calculus and Newton's law of gravity, Kepler's laws can be found very quickly.

Although Newton was the first to discover the mathematics of calculus, he was reluctant to publish his work. As a result, no one learned of his discovery until much later. Meanwhile, German mathematician and philosopher Gottfried Wilhelm Liebniz (1646–1716) independently discovered calculus and is usually listed as the codiscoverer with Newton. The notation for differentiation now used in calculus (∫) was put forward by Liebniz. Calculus is an extremely useful tool. It is used to figure out the orbits of planets and satellites, weather patterns, and ocean currents, for example, and even to make calculations in economic and sociologic theories.

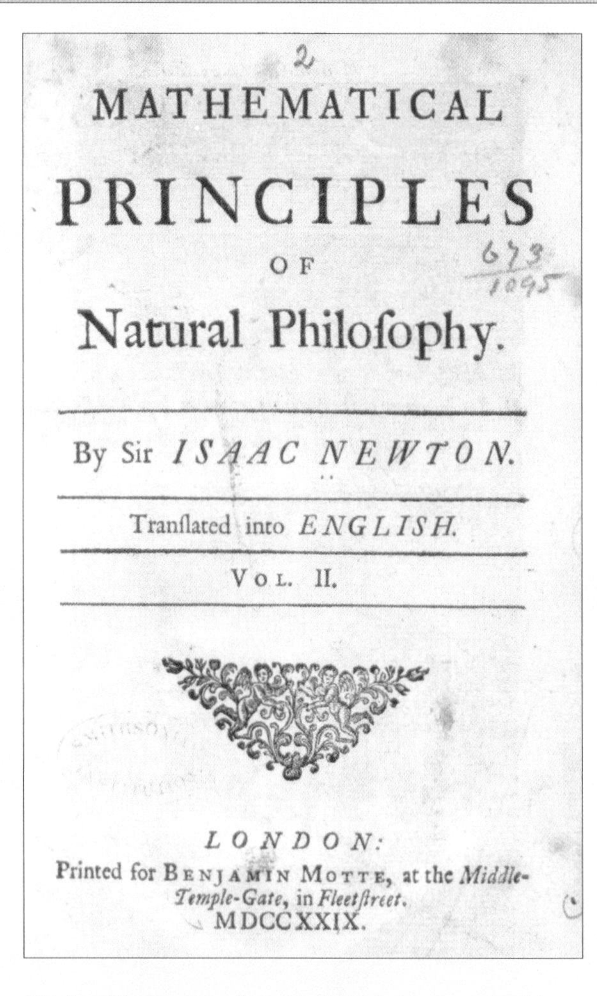

MATHEMATICAL PRINCIPLES OF Natural Philofophy.

By Sir ISAAC NEWTON.

Tranflated into ENGLISH.

VOL. II.

LONDON:
Printed for BENJAMIN MOTTE, at the Middle-Temple-Gate, in Fleetftreet.
MDCCXXIX.

Much of Newton's mathematical work was contained in* Principia*, originally written in Latin.

A CLOSER LOOK

Principia led Newton to a nervous breakdown, and his health was never the same again.

Politics and Parliament

A few years later, Newton became interested in politics and was elected to Parliament from Cambridge in 1689. In 1696 he was appointed warden of the Royal Mint and became involved with changing the British currency and catching counterfeiters. However, the appointment meant he had little time left to pursue his scientific activities.

With Hooke's death in 1703, Newton was elected president of the Royal Society of London. The following year, he published *Opticks*, which contained his work in the field from 30 years earlier. Two years later, in 1705, Newton was knighted by Queen Anne; he was the first scientist to be honored in such a way for his work. Isaac Newton died on March 20, 1727—or March 31, according to the Gregorian calendar—in London, at the age of 84. He was buried in London's Westminster Abbey.

C. KEATING/L. CAMPBELL-WRIGHT

See also: ACCELERATION; ASTRONOMY; CALCULUS; COLOR; COPERNICUS, NICOLAUS; COSMOLOGY; EINSTEIN, ALBERT; FORCES; GALILEI, GALILEO; GRAVITY; LIGHT; MOTION; NEWTONIAN PHYSICS; OPTICS; PHYSICS; RELATIVITY; SCIENTIFIC METHODOLOGY; SPACE EXPLORATION; SPACE SCIENCE; TELESCOPES; UNIVERSE.

Further reading:

Aughton, Peter. 2004. *Newton's Apple: Isaac Newton and the English Scientific Renaissance.* London: Weidenfeld and Nicolson.

Berlinski, David. 2002. *Newton's Gift: How Sir Isaac Newton Unlocked the System of the World.* Columbus, Ohio: Free Press.

Cohen, Bernard, and Anne Whitman, trans. 1999. *The Principia: Mathematical Principles of Natural Philosophy* by Isaac Newton. Berkeley, Calif.: University of California Press.

Crowther, J. G. 1995. *Six Great Scientists.* New York: Barnes and Noble Books.

Gleick, James. 2003. *Isaac Newton.* London: Random House/Pantheon Books.

NEWTONIAN PHYSICS

Newtonian physics is the body of theory consistent with Newton's laws of motion and gravitation

The term *Newtonian physics* (or classical physics) is used in contradistinction to *quantum physics* to denote the body of physical theory that is consistent with the laws of motion (see MOTION) and gravitation (see GRAVITY) first outlined by English scientist Isaac Newton (1642–1727; see NEWTON, ISAAC). The stars and planets, spacecraft, bullets, billiard balls, and even much smaller bodies all obey these laws to extremely high accuracy. They are not applicable to atoms and subatomic particles, however, which cannot be observed without significantly disturbing them (see QUANTUM THEORY).

Newton's laws were so successful in predicting planetary movements—and, as a result, detecting the existence of two other planets, Uranus and Neptune—that by the middle of the 19th century many scientists regarded the whole universe as working like a huge machine. The tiniest movement at one point, they argued, would inevitably be reflected in a predictable adjustment of the mechanical equilibrium of the whole. Toward the end of the same century, scientists declared that there was nothing more to be discovered in the physical world.

Then came the discovery of the electron and, within a generation, the proton and neutron (see SUBATOMIC STRUCTURE). In attempting to explain their behavior, physicists and mathematicians were compelled to accept inherent limits on their ability to describe the motion of physical objects.

One of the influences on this new way of thought was German-born U.S. physicist Albert Einstein (1879–1955; see EINSTEIN, ALBERT), who has been described as the last classical physicist. His theory of relativity was a limited refinement of Newtonian physics (see RELATIVITY). Einstein's equation of the relationship of energy with mass was essential to an understanding of the newly discovered particles, as was his elaboration of the work of German physicist Max Planck (1858–1947). Planck abandoned classical principles and in 1890 articulated the earliest forerunner to the quantum theory, without understanding all its implications. His theory successfully explained and predicted phenomena that could not be explained by Newton's. Other scientists developed the theory over the next 20 years, until quantum mechanics pulled everything together.

In the 21st century, Newtonian and quantum physics are of equal importance. The effects of quantum physics can be observed only on a subatomic scale or in many phenomena that are based on the behavior of large numbers of small particles. The dynamics of larger bodies can be explained with accuracy by Newtonian theory. However, objects moving at very high speeds, whether they are large or subatomic, also require the theory of relativity to be understood properly (see RELATIVITY).

Isaac Newton made tremendous contributions to many areas of the natural sciences.

CORE FACTS

- Newtonian, or classical, physics uses Newton's laws of motion and gravitation to explain and predict the movement of solid bodies.
- Quantum physics, by contrast, explains the behavior of atoms and subatomic particles.
- Newton's laws require modification, according to Einstein's theory of relativity, for bodies moving at or near the speed of light.
- Newton's great work, *Principia*, was written in less than two years and contains almost all of his discoveries and insights into the motion of objects.

Isaac Newton

Considered to be the single most important figure in the scientific revolution of the 17th century, Newton made brilliant contributions to many areas

CONNECTIONS

- The **FORCE** of **GRAVITY** is responsible for the elliptical orbits of planets in the **SOLAR SYSTEM**.

The force of gravity causes orbiting satellites to "fall" out of their natural straight-line path and toward Earth.

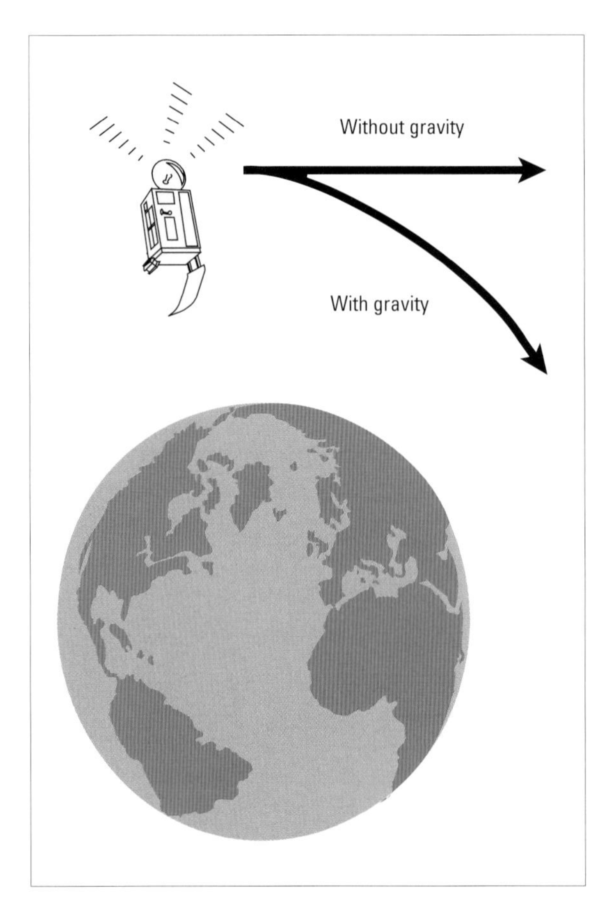

of the natural sciences. He is best known for his three laws of motion and his formulation of the law of universal gravitation, all of which he detailed in his 1687 publication *Philosophiae Naturalis Principia Mathematica* (Mathematical Principles of Natural Philosophy), usually called *Principia*. Among Newton's other achievements was his publication of *Opticks*, a comprehensive discussion of light and color, in 1704. Newton also developed calculus, an invaluable mathematical tool for his studies of motion and forces (see CALCULUS; MATHEMATICS). Building on the work of such scientists as Copernicus, Kepler, and Galileo, Newton created a powerful model for the understanding of the physical world, one that is still in use today (see COPERNICUS, NICOLAUS; GALILEI, GALILEO).

DISPUTES WITH OTHER SCIENTISTS

Newton's brilliant scientific career was marked by bitter contention with his fellow scientists, most notably with English scientist Robert Hooke (1635–1703). Newton's first scuffle with Hooke was after publication of Newton's theory of color and light in 1672. Hooke and others attacked Newton's ideas so relentlessly that Newton wrote in a letter to the secretary of the Royal Society, "I see a man must either resolve to put out nothing new, or to become a slave to defend it."

Later, after Newton had published the first two books of his *Principia*, Hooke demanded credit for ideas he said were mainly his own. Admitting to have been influenced by past thinkers, Newton responded, "If I have seen further it is by standing on the shoulders of giants." This statement was also a barbed comment directed at Hooke, who was a very short man. Newton went on to delete any reference to Hooke from the final book of *Principia* and delayed publication of his *Opticks* until after his rival's death.

Another source of controversy in Newton's life was his battle with German mathematician Gottfried Wilhelm Leibniz (1646–1716). At issue was which of them would take credit for formulating calculus. Leibniz published papers on calculus in 1684, and although his were the first published accounts of the method, Newton's unpublished work on calculus had taken place 20 years earlier.

HISTORY OF SCIENCE

Dynamics

The culmination of Newton's work in dynamics was *Principia*. This masterpiece, though written in an almost miraculous 18 months, was preceded by at least 20 years of thought and discussion on the nature of moving bodies. The first of the three books of *Principia* contains three laws, now famous as Newton's laws of motion: (1) A body continues in a state of rest or of uniform motion in a straight line unless compelled to change by forces acting on it (see FORCES). (2) The acceleration of an object is directly proportional to the total force acting on it and is inversely proportional to its mass (see ACCELERATION). The direction of the acceleration is in the direction of the applied force. (3) For every action there is an equal and opposite reaction.

The second book of *Principia* discusses the motion of bodies moving through resisting media (fluids, including gases and liquids) and the motion of fluids themselves. In this book, Newton demonstrates that the theory of French philosopher René Descartes (1596–1650), that the planets are kept in their orbits by swirling vortices of tiny particles, is impossible as well as inconsistent with the observations of German astronomer Johannes Kepler (1571–1630) on the motions of the planets.

Newton's law of universal attraction is given in the third book of the *Principia*, subtitled "System of the World." Newton's law of gravitation states that any two bodies will experience a mutual attractive force (F) that is proportional to the product of their masses (M and m) and inversely proportional to the square of the distance between them (R^2). Mathematically, the law of gravitation is written as follows:

$$F = G(Mm)/R^2$$

The gravitational constant, G, depends on the units used to measure mass and distance. G is a very small number, since the force of gravity is small unless the objects involved are very large.

Newton used his laws of motion and gravitation to explain the motion of the planets and their moons, the eccentric orbits of comets, the tides, and the equinoxes. As promised by its title, "System of the World" united the marvels of the universe under one simple set of laws. These laws reduced the cosmos to such an elegant mathematical system that Newton's work was widely regarded as proof of a divine creator. However, it took at least 50 years before scientists and universities would relinquish Descartes's vortex theory for Newton's more practical explanation of planetary motion.

It is important to remember that no scientist works alone. Newton's monumental work was a fusion of the work of the scientists who came before him as well as the work of his contemporaries. For example, without the laws put forth by Kepler 70 years earlier detailing the motions of the planets, Newton would probably not have developed his laws explaining planetary motion.

Likewise, Italian mathematician, astronomer, and physicist Galileo Galilei (1564–1642) did extensive study on the motion of moving bodies that contributed directly to Newton's work. Galileo was the first to observe that it is not necessary to constantly apply a force to a body to keep it moving. He concluded that rest and motion are both natural states, an idea that would later become Newton's first law of motion.

The merit of Newton's work in mechanics can be measured in terms of its longevity. Newtonian mechanics is still the foundation of the physical sciences, and it remains virtually unchanged over three centuries. However, Einstein's relativity theory has shown that, at speeds approaching the speed of light, Newton's laws no longer apply without modification. Although it is true that Newtonian mechanics does not hold for very fast or very small bodies, it still applies in most cases and is therefore an exceptionally useful tool.

Optics

One of Newton's most notable contributions to the understanding of light came as the result of experiments he conducted in the years 1661 to 1665. In an experiment that he dubbed "the crucial experiment," Newton showed that white light is not a purely homogeneous entity, as had been thought since the time of Aristotle. In this experiment, Newton passed a beam of sunlight through a prism in a darkened room, breaking it into an oblong spectrum that fell on a board. A small hole in this board allowed light of only one color to pass through to a second prism and, from there, on to a second board. Newton noted that while the white light from the Sun could be bent by the first prism into a spectrum of colors, a single color from this spectrum passed through a second prism would not decompose further (see LIGHT; OPTICS).

From this experiment, Newton concluded that white light is composed of a mixture of different types of rays, which are bent at different angles by the prism, and that each type of ray has its own color (see REFLECTION AND REFRACTION). He went on to say that these colored rays were "original," meaning that they could not be broken into other types of light. Scientists refer to this light as being monochromatic, literally meaning "one color."

Newton published his new theory of light and color in 1672. Although his theory was well received, Newton was also criticized by other scientists for claiming to have proven by experiment that light is composed of particles. At the time, it was widely held that light, like sound, is a wave, but Newton believed that the sharp shadows cast by light were evidence of light particles, since waves were known to bend around obstacles in their path (see DIFFRACTION AND INTERFERENCE). However, certain other light phenomena were difficult to explain except in terms of waves. In 1704 when Newton published *Opticks*, he clung to the particle theory but also incorporated aspects of a light-wave theory.

NEWTON AND THE FALLING APPLE

Legend has it that Newton recognized the force of gravity in 1666 while watching an apple fall from a tree in his garden at Woolsthorpe—and he himself later gave support to the tale. In fact, this story may not be entirely accurate, but it does illustrate Newton's revelatory breakthrough. Newton realized that the gravitational pull of Earth extends to the Moon. He reasoned that if gravity could be felt up in a tree or even on a mountaintop, it might reach as far as the Moon. This pull, Newton supposed, might be the force that keeps the Moon in its orbit, counterbalancing the natural tendency of an object to move in a straight line. Of course, it is now known that Newton's speculation was correct. What Newton did not propose until 20 years later is that gravity is a universal property and that every body exerts a gravitational pull that is proportional to its mass.

A CLOSER LOOK

Establishment of the scientific method

Newton's emphasis on mathematics and experiment set the standard for scientific method (see SCIENTIFIC METHODOLOGY). Previously, most scientific study had

White light was thought to be monochromatic until, in what Newton dubbed the "crucial experiment," he observed that white light from the Sun could be bent by a prism into the spectrum of colors shown above.

involved a high degree of conjecture with relatively little experimentation. Newton started work on his ideas with known facts and previous experimental data, formed a theory to explain the data in mathematical terms, and tested the natural conclusions of the theory against experimental fact, either already available or gathered from new experiments suggested by the theory.

Such an approach, relying so much as it did on definitive experimental and mathematical proof to support theories, became a powerful tool for scientific investigation, one that influenced many other scientific disciplines, including biology and analytical chemistry (see ANALYTICAL CHEMISTRY). Furthermore, Newton's development of new mathematical techniques, particularly differential and integral calculus, gave science a more exact tool for the expression of rapidly changing events.

P. TESLER

See also: ACCELERATION; ANALYTICAL CHEMISTRY; CALCULUS; COPERNICUS, NICOLAUS; DIFFRACTION AND INTERFERENCE; EINSTEIN, ALBERT; GALILEI, GALILEO; LIGHT; MASS; MATHEMATICS; MATTER; MECHANICS; MOMENTUM; MOTION; NEWTON, ISAAC; OPTICS; PHYSICS; QUANTUM THEORY; REFLECTION AND REFRACTION; RELATIVITY; SCIENTIFIC METHODOLOGY; SUBATOMIC STRUCTURE; WAVE MECHANICS.

Further reading:

Giambattista, Alan, Betty McCarthy Richardson, Robert C. Richardson. 2004. *College Physics* Boston: McGraw-Hill.

Roberts, Jeremy. 2001. *How Do We Know the Laws of Motion?* New York: Rosen Publishing Group.

NEWTON'S LIFE

Isaac Newton was born on December 25, 1642, in Woolsthorpe, Lincolnshire, England. His father died before his birth, and when he was two years old, his mother remarried and left him in the care of his grandmother. A quiet child, Newton was reportedly clever with his hands but inattentive at school.

In 1653 Newton's mother returned to Woolsthorpe after the death of her second husband, resolving to make a farmer out of young Isaac. After proving himself a miserable farmer, Isaac was allowed to return to school in preparation for entrance to Cambridge University.

Newton's performance at Cambridge caught the attention of his teachers, and he also spent much of his time engrossed in private study. When plague forced the university to close in 1665, Newton returned to Woolsthorpe, where he spent the next two years developing his "method of fluxions" (see CALCULUS; MATHEMATICS). At the same time, he performed optics experiments that would found his new theory of light and color, while also delving into the problem of planetary motion. Newton considered this two-year period his most productive.

He returned to Cambridge in 1667, earned his master of arts degree in 1668 and was offered a professorship. Soon after, he published his first paper, based on his earlier optical experiments. The controversy that this paper sparked—in particular, unfounded accusations that he had stolen ideas from his rival Robert Hooke—so disturbed Newton that he withdrew from society and suffered the first of several emotional breakdowns. In seclusion, he turned his attention to alchemical research.

In 1684, prompted by the British astronomer Edmond Halley (1656–1742), Newton resumed work on the problem of planetary orbits, a topic of great scientific interest at the time. In less than two years of intense effort, he produced *Principia*, which contained the three laws of motion and the law of universal gravitation. He was elected to represent Cambridge in Parliament in 1689, but after suffering another nervous breakdown, he left Cambridge without regret to take a government position at the Royal Mint in London.

In 1703 Newton was elected president of the Royal Society (the distinguished British scientific society), and shortly afterwards he published *Opticks*, based on the work he had done decades before. In 1705 he was knighted by Queen Anne. Although Newton's creative years were behind him, he wielded considerable influence as president of the Royal Society. He was reelected to this position annually until his death in London on March 20, 1727. Isaac Newton was buried in Westminster Abbey, London, the first scientist to be honored in this way.

DISCOVERERS

NICKEL AND COBALT

Nickel and cobalt are transition metals that are extensively used in alloys

Cobalt is a light-gray, ferromagnetic transition metal.

Neighbors in the periodic table and members of the nine groups of transition metals (see TRANSITION ELEMENTS), nickel (group 10) and cobalt (group 9) share many characteristics with each other and with iron (group 8; see IRON AND STEEL). These three metals have similar tensile strengths, thermal properties, and electrochemical behavior. Like iron, nickel and cobalt are the only elements that are magnetic at room temperature (see MAGNETISM).

It had been known in China for many centuries that cobalt ores would color porcelain blue. This effect was discovered in Europe in 1540, when a blue glass called smalt was manufactured. In 1730 Swedish chemist Georg Brandt (1694–1768) showed that the color was due to a previously undiscovered element, cobalt. In 1751 another Swedish chemist, Baron Axel Cronstedt (1722–1765), isolated nickel.

Chemical properties of nickel and cobalt

Like other members of the group 9 and 10 transition metals, nickel and cobalt are not very reactive. At room temperature, neither reacts readily with oxygen. However, the reaction progresses more rapidly at elevated temperatures. The most common oxidation states of nickel are +2 and +4, although nickel is also found in the 0, +1, and +3 states (but only rarely). The most common oxidation states of cobalt are +2 and +3, although −1, 0, +1, and +4 states exist (see OXIDATION-REDUCTION REACTIONS).

The most stable oxidation state of cobalt is Co(II). With water, it forms the complex ion $[Co(H_2O)_6]^{2+}$, which is deep pink (see IONS AND RADICALS). Crystals of cobalt chloride have the composition $[Co(H_2O)_6]Cl_2$ and are red, but when they are heated and the water is driven off, they turn blue. This property is sometimes used as a test for water in other liquids. Nickel forms a complex in the same way with water, but it turns emerald green.

From ore to metal

Both nickel and cobalt are found in Earth's crust. Nickel makes up 0.008 percent; cobalt represents 0.003 percent. Nickel, however, is thought to be a

CORE FACTS

- Nickel and cobalt are transition metals, belonging to periodic table groups 10 and 9, respectively.
- Nickel and cobalt are both ferromagnetic at room temperature.
- The Mond process is used to produce a highly pure form of nickel.
- Nickel is used extensively as an alloy and also in the manufacture of nickel-cadmium rechargeable batteries.
- Alloys of cobalt are used to produce permanent magnets.

NICKEL
Symbol: Ni
Atomic number: 28
Atomic mass: 58.69
Isotopes:
58 (68.27 percent),
60 (26.10 percent),
61 (1.13 percent),
62 (3.59 percent),
64 (0.91 percent)
Electronic shell structure: $[Ar]3d^84s^2$

COBALT
Symbol: Co
Atomic number: 27
Atomic mass: 58.93320
Isotopes:
59 (100 percent)
Electronic shell structure: $[Ar]3d^74s^2$

CONNECTIONS

- **ELECTROLYSIS** can extract **METALS** from their **ORES**.

COBALT AND VITAMIN B_{12} DEFICIENCY

In addition to the abundance of such substances as carbon, hydrogen, oxygen, and nitrogen that are necessary to sustain life, plants and animals also need the so-called trace elements. Found in living things in minute quantities, these elements are nevertheless essential for sustaining life.

Cobalt lies at the core of the molecule of vitamin B_{12}, which is essential for the production of red blood cells in animals. A deficiency of this vitamin causes a potentially serious blood condition called pernicious anemia. In mammals, the vitamin is not ingested whole but is synthesized by bacteria that are found in the digestive tract. The synthesis, however, cannot proceed without an adequate supply of cobalt, which must be taken in with food.

Cobalt deficiency in domestic animals, such as cattle and sheep allowed to graze on grass grown in soil deficient in cobalt, can cause a disease commonly called staggers, which has been prevalent in certain parts of southern Australia. The disease affects the central nervous system, and it can result in death. It can be prevented by adding small quantities of cobalt, usually as cobalt chloride or cobalt nitrate, to livestock feeds and fertilizers.

A CLOSER LOOK

major constituent of Earth's core. Both metals are found in meteorites (see METEORS AND METEORITES).

The leading producers of nickel are Russia, Canada, Australia, New Caledonia, and Indonesia, while the richest sources of cobalt lie in Zaire, Zambia, and Russia. Although the United States consumes about one-third of the world's cobalt production, its reserves of the metal are low grade, and it has not produced cobalt since 1971. Nickel, however, is produced in the United States, but the country imports a quantity equal to more than 60 percent of its needs. Most comes from the Sudbury region of Ontario, where two ores, the iron-nickel sulfides pentlandite and pyrrhotite, are mined (see MINING AND PROSPECTING). In Zaire, the world's major cobalt producer, the chief ores of cobalt are the oxide heterogenite and the sulfides carrollite and cattierite.

In a process called flotation, nickel-iron sulfide ores are crushed and mixed with water; when air is bubbled through, the nickel-rich material is carried off by the bubbles. (Some ores are magnetic and can be separated by magnetic means.) The concentrated ore is then smelted in a blast furnace, and an iron-nickel sulfide matte is produced, which contains up to 75 percent nickel (see SMELTING). Various processes, including electrolysis, are used to obtain the pure metal from the matte (see ELECTROLYSIS). In the Mond process, carbon monoxide (CO) is passed through the heated matte. Nickel reacts to form nickel carbonyl ($Ni(CO)_4$), a gas that can be condensed and decomposed to release pure metal.

Cobalt ores, depending upon their composition (they are often found in association with nickel or copper), are treated in similar ways. The pure metal is extracted finally by electrolysis.

Five-cent coins, or nickels, are actually a mixture of copper and nickel. This design was released in 2004 to commemorate the Lewis and Clark expedition into the Louisiana Territory between 1804 and 1806.

Uses of nickel and cobalt

Half of all the nickel manufactured is used in alloys with iron, particularly in stainless steel. Much of the rest is alloyed with copper to make monel metal, which is very hard and strong and corrosion resistant. It is used to make specialized tools and propeller shafts for ships and boats. Nickel powder is used as a catalyst in the manufacture of margarine by hydrogenation of oils (see CATALYSTS).

Products can be given a hard and shiny, tarnish-resistant plating of nickel by electrolysis of nickel solutions. Nickel hydroxide makes up the cathode of the nickel-cadmium rechargeable battery. Nickel-hydrogen batteries are used on spacecraft.

A major proportion of the world production of cobalt goes into the manufacture of alnico permanent magnets that are made from alloys of aluminum, nickel, and cobalt. They are very hard and retain their powerful magnetism over long periods. Cobalt alloy steels are used in components for jet engines and gas turbines because they remain strong and rigid at high temperatures.

Compounds of cobalt, many of which are a rich blue, are used as pigments for glass, ceramics, and enamels and in artists' materials and inks. The sulfate, $CoSO_4$, is used in electroplating and as an agricultural additive (see the box above) and the acetate, $Co(CH_3COO)_2$, is used as a bleaching agent.

C. PROUJAN

See also: ALLOYS; BATTERIES; CATALYSTS; ELEMENTS; IONS AND RADICALS; IRON AND STEEL; MAGNETISM; METALLURGY; METEORS AND METEORITES; MINING AND PROSPECTING; OXIDATION-REDUCTION REACTIONS; SMELTING; TRANSITION ELEMENTS.

Further reading:

Emsley, John. 2001. *Nature's Building Blocks: An A–Z Guide to the Elements.* New York: Oxford University Press.

NITROGEN AND NITROGEN CYCLE

Nitrogen is a gaseous element that is exchanged between the atmosphere, living organisms, and Earth's crust

Nitrogen is a colorless, odorless, and tasteless gaseous element that makes up about 78 percent, by volume, of Earth's atmosphere. It was discovered in the 1770s by three scientists working independently. They were British botanist Daniel Rutherford (1749–1819), Swedish chemist Carl Scheele (1742–1786), and British chemist Joseph Priestley (1733–1804). Both Priestley and Scheele (in the course of experiments in which they discovered oxygen in the air) found that when all the oxygen in a jar had been used by combustion, another gas remained in which nothing would burn. Rutherford found that a mouse could not survive in a sealed jar once it had consumed all the oxygen.

French chemist Antoine-Laurent Lavoisier (1743–1794), who listed all the known elements in the late 18th century (see LAVOISIER, ANTOINE), gave this new element the name *azote*, which means "no life." Some years later French chemist Jean-Antoine Chaptal (1756–1832) introduced the name *nitrogen* in recognition of the fact that it was a constituent of niter—potassium nitrate (KNO_3), or saltpeter.

Nitrogen is an essential component of proteins, the vital constituents of plant and animal cells. However, although animals take in nitrogen with every breath, the body is unable to use it in its elemental form to build these necessary proteins. Nitrogen is obtained from food, in which the nitrogen has already been converted to a usable, or "fixed," form.

Physical and chemical properties

Nitrogen is the seventh element in the periodic table, included in group 15 (VA; see PERIODIC TABLE). Its outer electron shell has five electrons; it therefore normally completes the stable eight-electron configuration by the formation of three covalent bonds (see CHEMICAL BONDS; VALENCE). They can be three single bonds, as in ammonia (NH_3); a triple bond, as occurs naturally in gaseous nitrogen (N_2); or a structural (N=N) bond, as in the azo compounds (see DYES; FUNCTIONAL GROUPS). All five electrons are involved in nitro compounds. The fact that the triple bond of N_2 is very strong and the atoms are tightly packed accounts for the stability of elemental nitrogen.

CORE FACTS

- Nitrogen (N) is the seventh element of the periodic table and a member of group 15 (VA).
- Nitrogen is a colorless, odorless, and tasteless gas.
- Nitrogen is an essential component in life processes because it builds cell proteins.
- Ammonia (NH_3) and nitric acid (HNO_3) are important commercial starting points for many essential compounds.
- Bubbles of nitrogen in the blood cause the bends, a potentially fatal disorder that can afflict deep-sea divers.

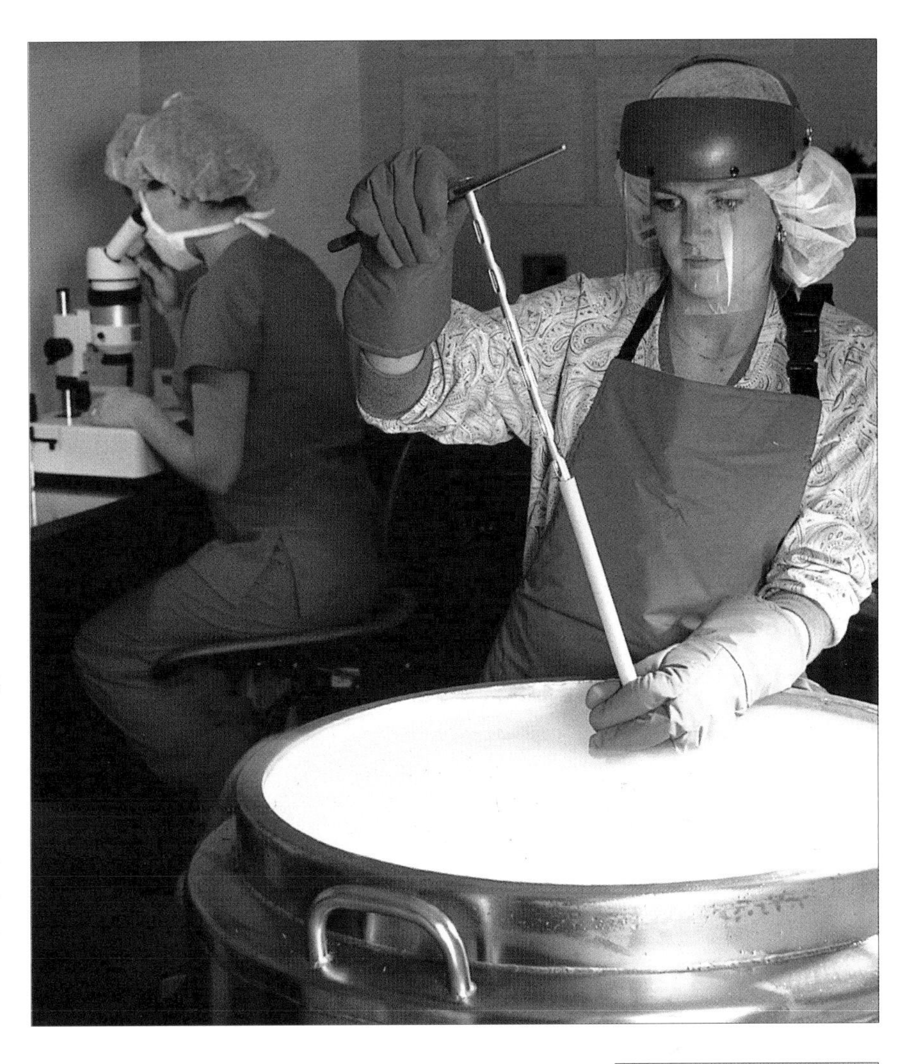

A technician removes samples of frozen sperm cells from a flask of liquid nitrogen. The sperm cells are used for in vitro fertilization (IVF) treatment.

Nitrogen gas is only slightly soluble in water, but it dissolves easily in liquid oxygen. When it is cooled under high pressure, it forms a colorless liquid with a boiling point of −320.8°F (−195.8°C). If the pressure is then rapidly reduced, further cooling takes place, with the formation of a colorless solid with a melting point of −346°F (−210°C).

Nitrogen can be prepared industrially from air by fractional distillation, which involves the evaporation of liquid air: liquid nitrogen boils at a higher temperature than oxygen (see MELTING AND BOILING POINTS). Dust particles, water vapor, and carbon dioxide are removed from the air, which is then liquefied at low temperatures and high pressures. Finally the nitrogen, oxygen, and argon are separated by distillation. Nitrogen gas can also be prepared by the decomposition of compounds of ammonium (NH_4^+) ions.

At temperatures below 392°F (200°C), nitrogen becomes unreactive. At higher temperatures it forms nitrides with many metals. These nitrides are easily decomposed by water to produce ammonia and the

CONNECTIONS

- **PURINES AND PYRIMIDINES** both contain nitrogen and are vital components of **NUCLEIC ACIDS**.
- **AMINES** are organic **COMPOUNDS** that are derived from ammonia.

Symbol: N
Atomic number: 7
Atomic mass: 14.00674
Isotopes:
14 (99.63 percent),
15 (0.37 percent)
Electronic shell structure: $[He]2s^22p^3$

corresponding metal hydroxide. At much higher temperatures, it combines with oxygen to form nitric oxide (NO). At very high pressures, it will combine with hydrogen to form ammonia (NH_3).

Besides its abundance in Earth's atmosphere, nitrogen is found naturally in many organic compounds, particularly plant and animal proteins. Many more nitrogen-containing organic compounds have been synthesized. The only major mineral source is Chile saltpeter, sodium nitrate ($NaNO_3$).

AMMONIA MANUFACTURE

Once scientists understood how nitrogen became fixed in plant growth, they began to devise ways to manufacture ammonia from atmospheric nitrogen. The first step in the process was the discovery by German chemist Carl Bosch (1874–1940) that hydrogen could be made by passing steam over white-hot coke, producing "water gas":

$$C + H_2O \longrightarrow H_2 + CO$$

In 1909 German chemist Fritz Haber (1868–1934), along with Bosch, developed the chemical process for industrial nitrogen fixation called the Haber-Bosch process, for which he earned the Nobel Prize for chemistry in 1918. Bosch, who scaled up the laboratory process so that it was industrially and commercially feasible, won the prize in 1931. In the Haber-Bosch process, hydrogen and nitrogen are passed through an iron oxide (Fe_2O_3) catalyst at a temperature of about 932°F (500°C) and a pressure as high as 1,000 bar:

$$3H_2 + N_2 \longrightarrow 2NH_3$$

The high pressures necessary for this process required a development of engineering techniques that were not possible before the 20th century. Today natural gas is often used as the source of hydrogen.

German chemist Fritz Haber was partly responsible for developing the Haber-Bosch process—the chemical process for industrial nitrogen fixation.

SCIENCE AND SOCIETY

Uses of nitrogen

As nitrogen is so unreactive, the gas is used in industrial processes. Atmospheric nitrogen is used on a huge scale in the synthesis of ammonia and nitric acid, which are used to manufacture fertilizers and explosives. Nearly 30 million tons of N_2 are produced annually in the United States.

Liquid nitrogen is used to freeze and maintain tissues at a low temperature and to store cultures of bacteria and viruses (see CRYOGENICS). It has been suggested that liquid nitrogen could be used to cool superconducting electrical power lines.

Oxides of nitrogen

Chemists recognize six oxides of nitrogen; nitrous oxide (N_2O), nitric oxide (NO), dinitrogen tetroxide (N_2O_4), nitrogen dioxide (NO_2), and dinitrogen pentoxide (N_2O_5). The most important inorganic nitrogen compounds are the acids and salts of the nitrite (NO_2^-) and nitrate (NO_3^-) radicals.

Nitrous oxide was first discovered by Joseph Priestley and was later investigated by British chemist Humphry Davy (1778–1829). This oxide has a sweetish smell, and since it is not poisonous, it can be breathed for some time when mixed with enough oxygen. Nitrous oxide was the first general anesthetic to be used. Davy inhaled, when he laughed and danced "like a madman," and ever since then it has been popularly known as laughing gas. Nitrous oxide is usually made by the decomposition of ammonium nitrate (NH_4NO_3).

Nitric oxide was also prepared by Priestley, who obtained it by reacting copper with nitric acid (HNO_3). It is a toxic gas, and its smell is unknown, because it immediately reacts with atmospheric oxygen to form the tetroxide N_2O_4. For industrial purposes, nitric oxide is produced through the catalytic conversion of ammonia (see CATALYSTS). It is also produced when the gases nitrogen and oxygen combine at elevated temperatures, as occurs in internal combustion engines or in a lightning flash. Nitric oxide is an important chemical messenger in the central nervous system and is released by white blood cells in the immune system as a chemical toxin.

Nitrogen dioxide and its dimer nitrogen tetroxide coexist in equilibrium as a brown gas with a pungent smell. This gas is very poisonous and is a major cause of air pollution by car exhausts (see POLLUTION).

Ammonia and nitric acid

Ammonia occurs naturally when organic nitrogen compounds, such as urea, are broken down by the action of bacteria. Animal manure, therefore, often smells characteristically of the gas. In small quantities ammonia is not very poisonous and stimulates the action of the heart. Smelling salts, a mixture of ammonium compounds that gave off the gas were used for this purpose. Liquid ammonia was previously used in refrigerators as a coolant, because its high latent heat of vaporization enabled it to carry off excess heat quickly. Since liquid ammonia is highly toxic, it was replaced with chlorofluorocarbons (CFCs). Domestic ammonia is a strong solution of the gas in water, as ammonium hydroxide (NH_4OH). Commercially synthesized ammonia is a starting point for many industrial products (see the box on page 1054).

Oxidation of ammonia by the Ostwald process is the usual method for the manufacture of nitric acid:

$$4NH_3 + 5O_2 \rightarrow 4NO + 6H_2O$$

A catalyst, such as platinum gauze, is needed for this reaction. The mixture of ammonia and air, which is passed through the gauze, needs to be heated at first to nearly 1832°F (1000°C), but since the reaction produces heat, the temperature reached is soon sufficient for the process to continue without additional heat. The nitric oxide is reacted with more air, and the gas is dissolved in water:

$$4NO + 3O_2 + 2H_2O \rightarrow 4HNO_3$$

Nitric acid is of great importance in industry. Much of it is combined with ammonia to make ammonium nitrate, both for use as a fertilizer and as a component of explosives (see EXPLOSIVES). Organic nitro compounds are also important as explosives and as products for the dye industry (see DYES).

The salts of nitric acid are called nitrates (see (SALTS). Nitrates are used in fertilizers and for making photographic chemicals (silver nitrate, $AgNO_3$) and red flares (strontium nitrate, $Sr(NO_3)_3$; see PHOTOGRAPHY).

Nitrites are used as food preservatives (sodium nitrite, $NaNO_2$; see HALL, LLOYD). Some scientists now believe that the body may convert nitrites to nitrosamines, which can cause cancer.

THE NITROGEN CYCLE

The nitrogen cycle is the way in which nitrogen circulates through the atmosphere, living organisms, and Earth's crust (see BIOSPHERE. It is essential to life on Earth. As nitrogen circulates through this cycle, there are changes in its oxidation state and energy, and it reacts with other elements such as oxygen, hydrogen, and carbon. The main processes involved in the nitrogen cycle include nitrogen fixation, in which atmospheric nitrogen is converted to ammonium ions (NH_4^+); nitrification, in which ammonium ions are converted to nitrates (NO_3^-); assimilation, in which nitrate is taken up or assimilated by life-forms; ammonification, in which nitrogen is excreted from life-forms in the form of ammonium ions produced by the decomposition of organic nitrogen; and denitrification, in which nitrates are converted back to nitrogen gas.

Nitrogen fixation

Breaking the covalent triple bonds of nitrogen gas requires an enormous amount of energy, so most organisms are unable to convert atmospheric nitrogen into a usable compound. However, blue-green algae and some other bacteria possess nitrogenase, a complex enzyme that facilitates the breaking of the bonds. Nitrogenase is irreversibly destroyed by oxygen, however, so the nitrogen-fixing process requires an anaerobic (oxygen-free) environment. The product of this process is the ammonium ion, NH_4^+.

Some nitrogen-fixing bacteria exist freely in the soil, whereas others occur in the oxygen-free root nodules of leguminous plants, such as peas, beans, and clover. These plants, therefore, are able to obtain their nitrogen indirectly from the soil via the bacteria.

Oxides of nitrogen are also formed directly in the atmosphere by lightning and are carried into the soil by rain. The nitrogen oxides then dissolve in the rainwater; the resulting nitric acid formed produces nitrates in the soil. It has been calculated that, worldwide, some 270,000 tons (245,000 metric tons) of nitric acid are formed in this way in the atmosphere each day.

This truck is spreading pig manure as a fertilizer. Like many of the waste products of living things, pig manure contains nitrogen compounds that can be used by plants to aid growth and development.

SMOG

Nitrogen dioxide (NO_2) is a chief constituent of smog, the dirty haze that envelops many cities. When organic materials such as unburned or partially burned fuels are added to the air, these organic molecules are oxidized into reactive fragments called free radicals (see IONS AND RADICALS).

Free radicals oxidize nitric oxide, and so reduce the concentration of this gas considerably. However, the result is a domino effect, in that ozone—which is generally used up in reactions with nitric oxide—rises in concentration eventually reacts with the organic molecules to speed up the overall production of free radicals. These react with a variety of atmospheric components to produce other smog components, such as aldehydes and ketones.

A CLOSER LOOK

HYDRAZINE

Hydrazine (N_2H_4) is derived from ammonia by reaction with sodium hypochlorite (NaOCl). Anhydrous hydrazine is a fuming colorless liquid that burns in air and generates a considerable amount of heat. These properties have made it a useful component of rocket fuels.

Hydrazine has other industrial uses as well. In power plants, for example, hydrazine is used to prevent corrosion of the metal parts of steam boilers by the oxygen that is dissolved in the water.

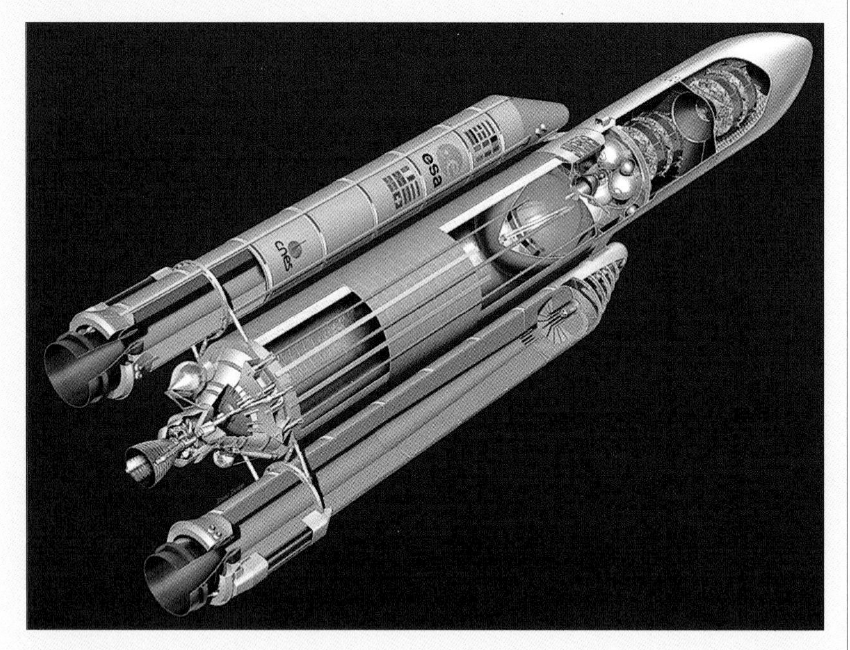

Nitrogen compounds are used to power spacecraft such as the European* Ariane 5*, shown above with its launcher.

Nitrification and assimilation

Although many plants can assimilate ammonium ions directly, it is easier for most of them to use nitrogen in the form of a nitrate ion. Nitrification is the conversion of ammonia to nitrate. It is performed by soil bacteria that obtain their energy by oxidizing compounds other than sugar. Nitrates are water soluble and are thus easily leached out of the soil into streams and groundwater (see GROUNDWATER; SOILS). For this reason, the soil does not store nitrogen in reserve, and plants require a continuous supply for growth.

Virtually all the ammonium ions formed in fixation are assimilated into organic nitrogen compounds, such as proteins (see PROTEINS) and nucleic acids (see NUCLEIC ACIDS) by living things.

THE BENDS

Decompression sickness, often called the bends, is a serious condition that results from the rapid change from high to normal atmospheric pressure. The bends often causes fainting and severe pains in the joints and muscles and in serious cases can lead to deafness, paralysis, and even death.

The bends develop because nitrogen dissolves in blood more readily under the higher pressures of deep-sea diving. If the pressure is reduced too quickly on returning to the surface, the respiratory system does not have sufficient time to gather and expel the excess nitrogen. Tiny bubbles appear in blood capillaries, and block the flow of blood and cause damage. To avoid the bends, scuba divers are advised to pace their ascent to the surface to give their body time to adjust; and deep-sea divers generally spend some time in decompression chambers, where the pressure is gradually reduced to the normal atmospheric level.

SCIENCE AND SOCIETY

Ammonification

Some nitrogenous material is returned to the soil as the excreta of animals. When plants and animals die, the nitrogen compounds then decompose to release ammonia, a process known as ammonification.

Denitrification

The nitrates are eventually reduced back to nitrogen by denitrifying bacteria that live in anaerobic environments. Oxygen is produced as a by-product, and the bacteria use it to obtain energy from carbohydrates.

Disruption of the nitrogen cycle

Because harvesting crops interrupts the natural process of decay, which would normally return nitrogen to the soil, farmers supplement natural sources of nitrogen with fertilizers that contain nitrogen. Most modern fertilizers are synthetic. On a global basis, approximately 90 million tons (82 million metric tons) of industrially produced nitrogen are added to the soil each year.

This heavy current fertilizer use and expected increase in use is of grave concern to many scientists. Without inputs of industrially produced fixed nitrogen, the global nitrogen cycle is stable; over time, approximately the same amount of nitrogen is fixed as is removed by denitrification. The industrially produced fixed nitrogen causes a major disruption of this important natural cycle. The buildup of nitrogen can produce several different ecological problems. Runoff from agricultural land can lead to excess concentrations of nitrates in streams, rivers, and lakes, contributing to a process called eutrophication. Algae that feed off nitrates grow and reproduce dramatically, blooms of algae being the result. These algal blooms lead to competition and shading, and eventually the algae die because they are unable to obtain sufficient nutrients and sunlight. As their cell materials decay through aerobic metabolism, oxygen is depleted from the water—oxygen that is critical for fish and other organisms.

S. FENNELL

See also: BIOSPHERE; CATALYSTS; CHEMICAL BONDS; CHEMICAL REACTIONS; CRYOGENICS; FUNCTIONAL GROUPS; GASES; GROUNDWATER; HALL, LLOYD; LAVOISIER, ANTOINE; POLLUTION; SOILS; VALENCE.

Further reading:

Brown, T. L. et al. 2002. *Chemistry: The Central Science.* Upper Saddle River, N.J.: Prentice-Hall.
Emsley, John. 2001. *Nature's Building Blocks: An A–Z Guide to the Elements.* New York: Oxford University Press.

NOAA

NOAA is a federal scientific agency that conducts research into the oceans and atmosphere

The National Oceanic and Atmospheric Administration (NOAA) was established in 1970 as a component of the U.S. Department of Commerce (DOC). NOAA combined the activities of some of the nation's oldest scientific agencies, including the National Ocean Service, established in 1807 to chart the nation's coastline; the National Marine Fisheries Service, established in 1871; and the National Weather Service, established in 1890. Following a decision by Congress in the late 1990s some NOAA programs were reduced in size and others were privatized. NOAA currently employs about 13,000 scientists, technicians, and administrators and is served by another 10,000 volunteers. The NOAA budget averages about $3 billion annually .

NOAA consists of five major agencies—the National Weather Service; the National Ocean Service; the National Marine Fisheries Service; the National Environmental Satellite, Data, and Information Service; and the Office of Oceanic and Atmospheric Research—plus a planning unit, several special programs, and the NOAA Corps, a uniformed service that operate NOAA vessels and aircraft.

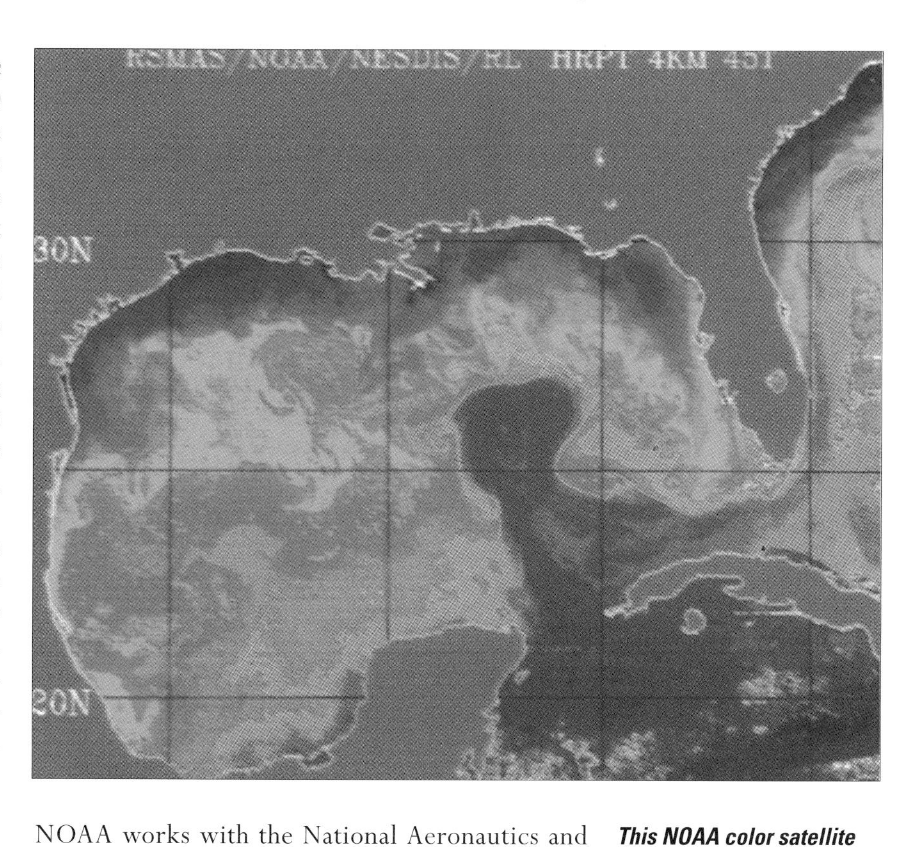

This NOAA color satellite image of the heat distribution in the waters of the Gulf of Mexico shows the Loop current (red) flowing around the island of Cuba (gray finger at bottom right).

Main themes of NOAA's work

NOAA conducts research and disseminates information in the following areas:

- Weather prediction, including day-to-day information used by media weather forecasters, pilots, and farmers and early warnings for severe weather events such as tornadoes, hurricanes, blizzards, and floods.
- Environmental research, especially to understand and perhaps help reverse environmental damage such as oil spills, the degradation and depletion of habitats in wetlands and estuaries, ozone depletion, the greenhouse effect, and possible global environmental change, including global warming (see GLOBAL WARMING; POLLUTION).
- Management and protection of marine plants and animals, including marine mammals, corals, sea turtles, and fish and shellfish species of value to recreational and commercial fishers.
- Basic research in oceanography, meteorology, biology, and physics.
- Cartography, including the production of air and sea charts.

NOAA coordinates its research and regulatory efforts with those of numerous local, state, federal, and international groups and agencies. For example, NOAA works with the National Aeronautics and Space Administration (NASA) to launch, maintain, and control satellites (see NASA); the International Whaling Commission to protect dolphins and whales; and groups, such as the International Halibut Commission, to manage commercial and recreational fisheries. NOAA also coordinates its work with that of the U.S. Fish and Wildlife Service in order to restore and protect wetland habitats, manage anadromous fish—those that migrate up rivers from the sea in order to breed, such as trout and salmon—and protect endangered species, including seabirds and sea turtles.

CORE FACTS

- NOAA is a federal agency that carries out research into the oceans and the atmosphere.
- NOAA data is used to produce general weather forecasts and to predict severe weather conditions.

Weather forecasting

The National Weather Service (NWS) provides short-term weather predictions and long-term climate outlooks. During the mid-1990s the NWS modernized and restructured its operations to take advantage of advances in radar, satellite, and information processing technology (see NATIONAL WEATHER SERVICE). Scientists analyze images from NOAA satellites to forecast the weather and to search for signs of hurricanes, tornadoes, and other severe weather events. NOAA satellite images are also used in TV weather reports (see SATELLITES, ARTIFICIAL).

Satellites and data management

NOAA's National Environmental Satellite Data and Information Service (NESDIS), established in the early 1980s, was formed by the merger of the former National Earth Satellite Service with the

CONNECTIONS

- NOAA **SATELLITES** are used to forecast severe weather events such as **FLOODS**, **HURRICANES**, and **TORNADOES**.
- **GLOBAL WARMING** and the depletion of the **OZONE LAYER** are monitored by NOAA.

Environmental Data and Information Service. NESDIS distributes information from NOAA satellites and from buoys operated by other agencies.

NOAA operates two polar-orbiting and two geostationary satellites. Geostationary satellites always remain fixed in the same spot high above the equator, because they move through space at the same rate that Earth rotates. They are positioned approximately 21,748 miles (35,000 km) above Earth's surface. NOAA's two Geostationary Operational Environmental Satellites (*GOES West* and *GOES East*) measure Earth-emitted and reflected radiation. This data can be used to derive information on atmospheric temperature, winds, moisture, and cloud cover. GOES satellites collect data from the central and eastern Pacific Ocean (including Hawaii and the Gulf of Alaska); North, Central, and South America; and the central and western Atlantic Ocean. *GOES West* hovers at 135 degrees west longitude, and *GOES East* at 75 degrees west longitude.

Polar-orbiting satellites circle Earth in strips from Pole to Pole and provide a continuously changing and detailed weather picture from much closer to Earth, at about 530 miles (853 km). Since the late 1990s the two polar satellites have held Sun-synchronous orbits, *NOAA-12* in an orbit at 500 miles (805 km) and *NOAA-14* in an orbit at 530 miles (853 km). *NOAA-12* was launched by NASA on May 14, 1991, and *NOAA-14* on December 30, 1994. The instruments on *NOAA-12* and *NOAA-14* measure Earth's surface and cloud cover, as well as solar protons, positive ions, and other atmospheric variables.

In addition, *NOAA-14, GOES West,* and *GOES East* receive and transmit data from search-and-rescue beacons, a vital service that directs coast guards and other rescuers to vessels and aircraft in distress. NOAA claims that in five years, this system helped save 5,448 lives.

NOAA satellites can forecast severe weather events, such as Hurricane Hugo, which left a trail of destruction across South and North Carolina in September 1989. This hurricane, with maximum wind speeds of around 160 miles per hour (260 km/h), was the strongest hurricane that year.

NOAA also supports 10 other satellites: *GOES-2, GOES-3, GOES-7, NOAA-9,* and *NOAA-11,* and some military weather satellites in the Defense Meteorological Satellite Program. NESDIS is also responsible for supporting the continued operation of existing LANDSAT (Land Remote Sensing) satellites (see REMOTE SENSING).

Protecting coastal areas

One of the main jobs of NOAA's National Ocean Service (NOS) is to protect coastal areas. These coastal areas include estuaries, which are areas where freshwater mixes with saltwater, such as many river mouths and bays. Estuaries are important nursery areas and sources of food for many types of fish and wildlife. To protect estuaries, NOS manages a 450,000-acre (182,000 ha) system of estuarine reserves in cooperation with state governments and local agencies. NOS works with local communities to prevent and minimize damage from oil and chemical spills and chronic contamination. It also maintains a national network of monitoring programs to detect, quantify, and forecast changes in the environmental quality of coastal areas.

NOS specialists work with state and local governments to develop regulations and management plans that balance many, sometimes competing, uses for coastal areas, including environmental protection, development (such as the construction of houses and hotels), transportation, and fisheries. For example, NOS is involved with various questions: Should owners of beach houses that have been destroyed by hurricanes be allowed to rebuild? Can a developer building a hotel destroy eelgrass beds that serve as nursery areas for fish stocks?

NOS also manages 14 National Marine Sanctuaries, which are protected undersea areas where certain activities, such as dredging are not allowed. Sanctuaries include a wide range of habitats and wildlife, from kelp beds, abalones, and gray whales off California's Channel Islands, to coral reefs and tropical fish in the Florida Keys, as well as historic shipwrecks. NOAA's challenge in managing these sanctuaries is to allow human use (such as recreational diving, kelp harvesting, and some types of commercial fishing) while providing secure habitats for species with depleted populations.

NOS also produces tidal and current tables and navigation charts for all U.S. coastal waters, including the Great Lakes and the Exclusive Economic Zones (EEZ) surrounding U.S. territories. NOS includes the Office of the National Geodetic Survey, which surveys and maps Earth's surface via satellite.

Regulating fisheries

Staff at NOAA's National Marine Fisheries Service (NMFS) study and help regulate a wide range of commercial and recreational fisheries, from haddock and lobster in Massachusetts to grouper in Florida and pollock and salmon in the North Pacific. NMFS scientists work with state governments and regional fishery management councils to set and regulate harvest levels. The goal of many of these regulations is to establish sustainable fisheries that can be maintained for many years without damaging fish stocks. NMFS has jurisdiction over many fish stocks in the EEZ of the United States, which extends as far as 200 nautical miles off the U.S. coast and thus is the world's largest EEZ.

In recent years NMFS scientists have concentrated their research efforts on finding ways to restore depleted fish stocks. For example, the Subcommittee on U.S. Coastal Ocean Science has determined that nearly half of U.S. fish stocks are being overexploited, including species such as Atlantic cod. To prevent any further decline, scientists in NMFS and other NOAA divisions are working with private and public landowners in an attempt to restore damaged marine areas that are important nursery and spawning grounds for fish.

NMFS works to protect endangered species such as marine mammals and sea turtles. NMFS also enforces fishery regulations; collects statistics on fishery catches, exports, and imports; inspects seafood plants; and advises U.S. companies that export their seafood products.

Researching the ocean and atmosphere

The goal of NOAA's Office of Oceanic and Atmospheric Research (OAR) is to study Earth as a system extending from the surface of the Sun to the ocean floor. Topics of study include the ozone layer, global warming, El Niño events (see EL NIÑO), fishery productivity, water quality, aquaculture technology, and ecosystem health. The research carried out by OAR supports the other branches of NOAA.

OAR operates environmental research laboratories on the Atlantic, Pacific, and Gulf coasts and conducts research projects jointly with many academic institutions and government agencies. OAR also funds and oversees work at National Sea Grant College Programs and Undersea Research Centers. Sea Grant Programs fund marine and freshwater research at more than 300 academic and research institutions; Sea Grant education and outreach programs make research results available to industry and local communities. The National Undersea Research Program develops technology to help scientists conduct undersea research. These tools include piloted submersibles, remotely operated underwater vehicles (including new models being developed by Sea Grant researchers at Woods Hole Oceanographic Institute and the Massachusetts Institute of Technology; MIT), and the undersea research station *Aquarius*, deployed in the Florida Keys (see MARINE EXPLORATION).

The NOAA fleets

The officers of the NOAA Corps and NOAA civilian personnel, which operate research vessels in the Atlantic and Pacific Oceans, conduct hydrographic surveys and fisheries and oceanographic research. They also pilot aircraft on missions to map coastal areas and improve hurricane prediction capabilities.

The NOAA Corps currently operates 18 ships and 10 aircraft, including two helicopters. The NOAA Corps is the United States' seventh and smallest uniformed service (with about 330 officers)—the other U.S. uniformed services are the U.S. Army, Navy, Air Force, Marine Corps, Coast Guard, and Public Health Service. Like other government programs such as NASA and the National Weather Service (see NASA; NATIONAL WEATHER SERVICE), NOAA programs are subject to fluctuations in size depending on government decisions about changes in funding.

K. FREEMAN

See also: ATMOSPHERE; CLIMATE; EL NIÑO; MARINE EXPLORATION; MESOSPHERE; METEOROLOGY; NASA; NATIONAL WEATHER SERVICE; OCEANS AND OCEANOGRAPHY; REMOTE SENSING; STRATOSPHERE; THERMOSPHERE; TROPOSPHERE; WEATHER.

Further reading:

Davies, N. 2004. *Oceans and Seas*. New York: Kingfisher.
Gallant, Roy A. 2003. *Atmosphere: Sea of Air*. New York: Benchmark Books.
Woodward, John. 2004. *Oceans*. North Mankato, Minn.: Smart Apple Media.

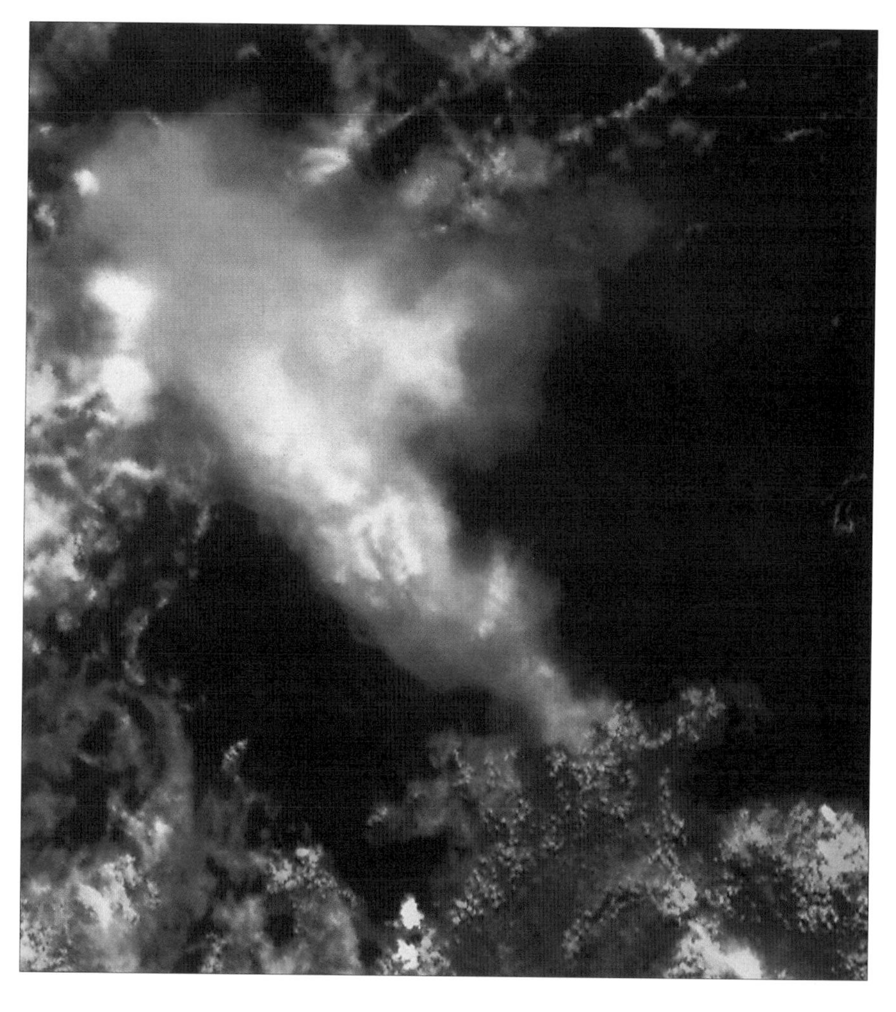

This satellite image, produced by the* NOAA-11 *polar-orbiting satellite on June 11, 1991, shows a cloud of volcanic ash rising from Mount Pinatubo in the Philippines soon after an explosive eruption.

NOBLE GASES

The noble gases are a group of unreactive elements at the right on the periodic table

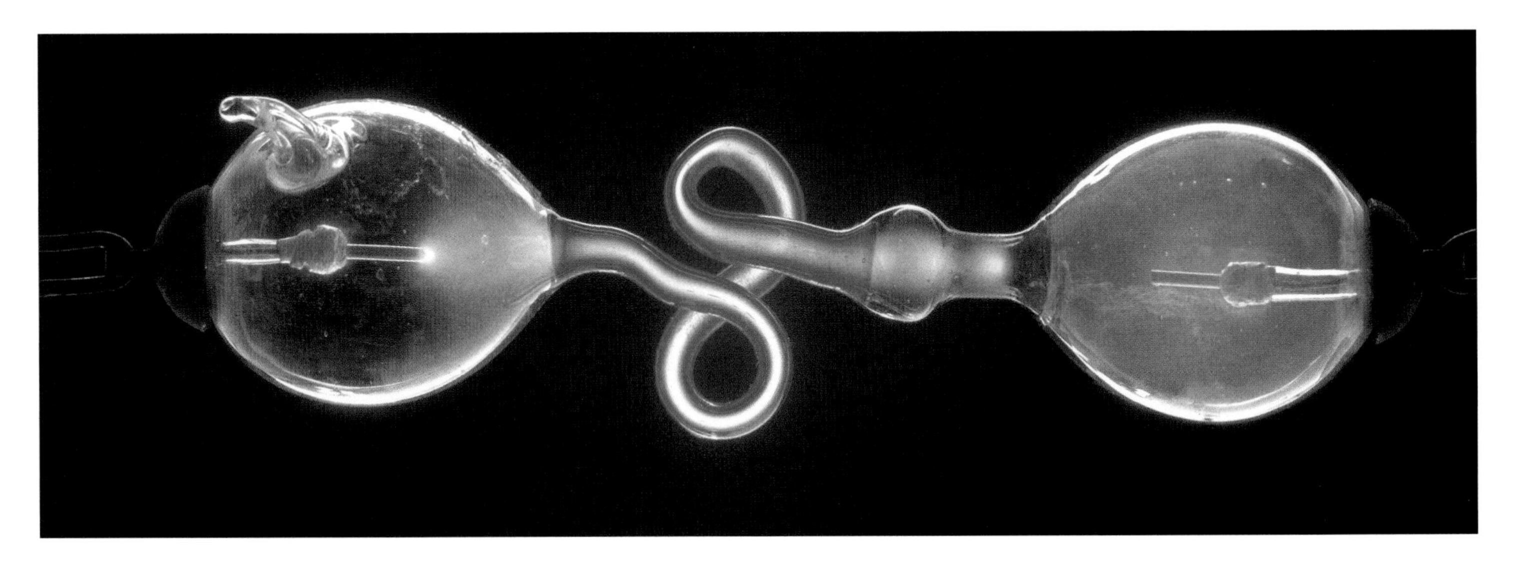

This glass discharge tube is a replica of the forerunner of modern neon lights, invented in the 1850s by German glass blower Heinrich Geissler (1814–1879). Neon lights are not always filled with neon gas. Many also contain argon. A high-voltage current is passed between two metal electrodes through argon gas and mercury vapor. The current causes the vapor to emit light.

At the extreme right of the periodic table of the elements, there exists a family of elements that are virtually unreactive. These elements are the so-called noble gases. The members of the family—helium, neon, argon, krypton, xenon, and radon—were dubbed "noble" because, like royalty, they shunned interactions with other members of their elemental community. The group is also called the inert gases.

The chemistry of the noble gases

The reason for the unreactivity of the noble gases lies in the configuration of their outer electron shells, which consist of a full complement: two in the case of helium and eight in the cases of the other noble gases. Since reactions between elements are driven to a large extent by the need for an atom to complete its outer electron shell by either capturing, relinquishing, or sharing some or all of these electrons, the complete shells of the noble gases make such reactions almost impossible. In other words, these elements have both high ionization energies (the energy needed to remove an electron from a gaseous atom in its ground state) and low electron affinities (the potential energy change in a gaseous atom following the addition of an electron; see ATOMS; ELECTRONEGATIVITY; VALENCE).

This behavior intrigued and challenged chemists, who toward the middle of the 20th century set out to produce artificial compounds of at least some of the noble gases. The focus of the chemists' work was on the heavier noble gases, since the nuclei of their atoms would exert less of an attraction for their outer electrons (being farther away from the nucleus) than would the nuclei of the lighter noble gases. They thought the heavier elements might react more readily because they had lower ionization energies. Thus, a powerful electron attractor—such as chlorine, fluorine, or oxygen or some complex substance containing them—might, under the right conditions, react with a heavy noble gas.

This hypothetical situation was turned into reality when in 1962 British chemist Neil Bartlett (b. 1932), working at the University of British Columbia, reacted a complex compound of fluorine—platinum

HELIUM
Symbol: He
Atomic number: 2
Atomic mass: 4.002602
Electronic shell structure: $1s^2$

NEON
Symbol: Ne
Atomic number: 10
Atomic mass: 20.1797
Electronic shell structure: $[He]2s^22p^6$

ARGON
Symbol: Ar
Atomic number: 18
Atomic mass: 39.948
Electronic shell structure: $[Ne]3s^23p^6$

KRYPTON
Symbol: Kr
Atomic number: 36
Atomic mass: 83.80
Electronic shell structure: $[Ar]3d^{10}4s^24p^6$

XENON
Symbol: Xe
Atomic number: 54
Atomic mass: 131.29
Electronic shell structure: $[Kr]4d^{10}5s^25p^6$

RADON
Symbol: Rn
Atomic number: 86
Atomic mass: 222
Electronic shell structure: $[Xe]4f^{14}5d^{10}6s^26p^6$

CONNECTIONS

- A **GAS** is a diffuse state of **MATTER** in which molecules have almost unrestricted **MOTION**.

CORE FACTS

- The noble gases belong to group 18 of the periodic table of elements.
- The noble gases are extremely unreactive, because their outer electron shells contain a full complement of electrons.
- Although compounds of noble gases do not exist in nature, chemists have been able to create artificial compounds of the heavier noble gases in the laboratory.
- The noble gases are used in welding, in fluorescent lights, and in balloons and blimps.

hexafluoride—with xenon to produce xenon platinofluoride ($XePtF_6$). It was soon followed by the synthesis of a variety of fluorine and oxygen compounds of xenon and a fluoride of krypton.

Discovering the noble gases

The first hint of the existence of the noble gases was provided by British chemist and physicist Henry Cavendish (1731–1810), who in 1785 discovered there was a component in air that would not react to form a compound. The second hint came with the development of the periodic table of the elements in 1869 by Russian chemist Dmitry Mendeleyev (1834–1907). Mendeleyev organized the 63 then known elements in order of atomic weights, a method that also produced periodic rises and falls of valences from 1 to 7 (see MENDELEYEV, DMITRY; PERIODIC TABLE). Notably absent were substances with a valence of 0. At that time, many chemists believed the reason for the absence was that no such substances existed.

This belief was laid to rest in 1894 when British chemist William Ramsay (1852–1916) and English physicist Lord Rayleigh (John William Strutt; 1842–1919) discovered argon in air. This gas was Cavendish's unreactive component in air. Ramsay, together with British chemist Morris William Travers (1872–1961), also discovered krypton, xenon, and neon in air, in 1898. In 1900 German physicist Friedrich Ernst Dorn (1848–1916) found that a product of radium decay was a previously undiscovered gas, which was later given the name radon (see RADIUM).

The story of the discovery of helium is fascinating because it was first discovered not on Earth but in the Sun. By the middle part of the 19th century, scientists were using an instrument called a spectroscope to identify chemical elements by the telltale lines they produced when heated (see SPECTROSCOPY). On August 18, 1868, a solar eclipse took place that afforded astronomers an opportunity to study the spectral lines of the Sun's fiery edge. To the surprise of no fewer than six observers, a unique line appeared in the spectrum of the Sun, one that did not agree with any produced by the then known elements.

For the next three years, scientists speculated about the significance of the mysterious line. In 1871 British astronomer Joseph Lockyer (1836–1920) reached the inescapable conclusion that the line was produced by a new element, which he named helium, after the Greek word for the Sun (see HELIUM). It was not until 1895 that, while studying the spectrum of a gas liberated from a mineral called cleveite, Ramsay discovered the same line that had been observed in 1868 in the Sun's spectrum. Helium did indeed exist on Earth. With these discoveries, group 18 (0) of the periodic table was complete.

Uses of the noble gases

The most obvious uses of the noble gases are in the production of colored light in fluorescent tubes. Neon produces red light when excited by electricity. Various ratios of argon and neon produce shades of blue or green. The major use of argon is in welding, where the gas is used to envelope the metal being welded so that oxidation is inhibited. Krypton lamps are extremely bright. For this reason, airplane runways are often illuminated with krypton-filled lamps that flash 40 times a minute. Such lamps can pierce dense fog for 1,000 feet (300 m) or more. Xenon-filled arc lamps are used in motion picture projectors. Radon, which is radioactive, has been used to determine the movement of air masses in Earth's atmosphere.

Helium has a wide variety of uses, the most well-known of which is in balloons and blimps. Its major use, however, is in welding, where it protects metals from oxidation while allowing for the production of maximum heat. The use of argon in this application reduces the excess production of heat.

C. PROUJAN

See also: ATOMS; ELECTRONEGATIVITY; ELEMENTS; HELIUM; MENDELEYEV, DMITRY; PERIODIC TABLE; RADIUM; SPECTROSCOPY; VALENCE.

Further reading:

Emsley, John. 2001. *Nature's Building Blocks: An A–Z Guide to the Elements.* New York: Oxford University Press.

In 1894 British chemist William Ramsay isolated the inert gas argon from air.

NONMETALS

Nonmetals are elements that, typically, tend to gain electrons in chemical reactions with metals

Sulfur is a yellow, brittle, crystalline nonmetallic element. It occurs naturally in its native form, as shown above, and in sulfide ores such as galena and zinc blende (sphalerite).

Of the 92 natural elements, only 17 are nonmetals. These are, in order of their atomic weights, hydrogen (H), helium (He), carbon (C), nitrogen (N), oxygen (O), fluorine (F), neon (Ne), phosphorus (P), sulfur (S), chlorine (Cl), argon (Ar), selenium (Se), bromine (Br), krypton (Kr), iodine (I), xenon (Xe), and the radioactive element radon (Rn).

The periodic table was originally put together on the basis of the chemical properties of the elements and their atomic weights (see ELEMENTS; PERIODIC TABLE). However, one previously unknown fact became immediately obvious. Generally speaking, in crossing the table from left to right in any one period, typical metallic properties gradually decreased, and in going down the table in any one group, metallic properties gradually increased. As knowledge of the atomic structure of the elements was assembled, it became clear that the position in the table also reflects the structure of the atoms.

The atomic structure of nonmetals

The nonmetals, which all have relatively low atomic masses, are placed in the upper right-hand corner of the periodic table. Hydrogen, the lightest and simplest of the elements, occupies a unique position. It has many of the chemical properties of a metal, but because of its physical properties (it is a gas, not a solid) and because it will also combine with metals, it is generally classed as a nonmetal.

The atomic number of an element is equal to the number of protons in its nucleus, and its atomic mass is a measure of the total number of protons and neutrons in the nucleus (see ATOMS). The protons give the nucleus a positive charge, and this charge is balanced by the negatively charged electrons that surround it, one electron for every proton (see ELECTRONS AND POSITRONS; NEUTRONS; PROTONS). These electrons are at increasing distances from the nucleus, in different energy levels, or "shells."

The first shell can contain no more than two electrons, the structure of helium (atomic number 2). The second shell can contain up to eight electrons. In period 2, the first nonmetal is carbon (6), which has four electrons in its second shell; after carbon come nitrogen (7), oxygen (8), fluorine (9), and then neon (10), which has the full number of electrons in both its shells.

In period 3, the third electron shell is built up in a similar way. Phosphorus (15) has five outer electrons, sulfur (16) six, chlorine (17) seven, and argon (18) eight. However, the total possible number of electrons in this shell is 18. In period 4, the nonmetals have the full number of 18 electrons in the third shell, and more electrons are added to the fourth shell. Selenium (34) has six, bromine (35) has seven, and krypton (36) has eight outer electrons.

The total number of electrons allowed in the fourth shell is 32, and this quantity results in a wide range of metals, as in period 4. Following the rule that metallic properties increase going down the periodic table, there are only two true nonmetals in period 5: iodine (53), with 7 electrons in the fifth shell; and xenon, (54) with 8. Finally, radon (86) has 32 electrons in the fourth shell, 18 in the fifth, and 8 in the sixth.

Properties of nonmetals

Nonmetals and metals generally have very different physical properties. At room temperature and pressure, all but five of the nonmetals (carbon, phosphorus, sulfur, bromine, and iodine) are gases. Bromine is the only nonmetal existing as a liquid under normal conditions. The physical property that most distinguishes nonmetals and metals is electrical conductivity: nonmetals do not readily give up electrons and so are nonconductive.

CONNECTIONS

- Twelve nonmetals are **GASES** at room **TEMPERATURE** and pressure.
- The **CHEMICAL REACTION** that results from the combination of the nonmetal **OXYGEN** and the metal **IRON** produces rust.

CORE FACTS

- In contrast to metals, nonmetals are poor conductors of electricity.
- Most nonmetals are gases at normal temperatures.
- Solid nonmetals are dull and brittle.
- Nonmetals tend to form ionic substances when they chemically combine with metals.
- Nonmetals chemically combine with other nonmetals to form covalent substances.

This flask contains crystals and purple vapor of the nonmetallic element iodine.

In reactions with metals, typical nonmetals tend to gain electrons given up by the metals. When nonmetals gain electrons, they become anions (negatively charged ions). The loss of electrons by metals produces cations (positively charged ions). The chemical compound formed in the reaction of nonmetals with metals is generally ionic (see CHEMICAL BONDS; IONS AND RADICALS). The force that holds cations and anions together in an ionic substance is particularly strong because of the naturally strong direct attraction between oppositely charged particles.

Nonmetals accept electrons to make up the full number in the outer shell. The gases helium, neon, argon, krypton, xenon, and radon, however, already have a full set of eight electrons, and the lighter members of this group do not react at all. At one time it was thought that none of these gases reacted; they were called noble gases. It is now known that under very special conditions, krypton, xenon, and radon can form compounds with fluorine, oxygen, chlorine, and nitrogen (see HALOGENS; NOBLE GASES).

Other nonmetals form compounds of this kind quite readily. When different nonmetals combine chemically, electrons are not transferred from one to the other. Instead atoms of the nonmetals share electrons, forming a covalent bond. As free elements, atoms of the nonmetals will also form covalent bonds with one another. The gases nitrogen, oxygen, fluorine, and chlorine—and bromine and iodine—form molecules made up of two atoms. Carbon is capable of forming huge molecules with chains and rings of covalent bonds (see CARBON).

Occurrence and abundance

The nonmetals hydrogen and helium are the two most abundant elements in the universe. Together these lightest of elements account for more than 99 percent of the atoms in the universe. On Earth the most abundant nonmetal is oxygen, which is found in the atmosphere as O_2, in the oceans as H_2O, and in Earth's crust as silicates (compounds containing silicon and oxygen) and carbonates (containing carbon and oxygen). The other nonmetals are found in smaller abundance, usually combined with other elements.

Six nonmetals—hydrogen, oxygen, nitrogen, carbon, sulfur, and phosphorus—form the basis for all biologically important molecules and account for more than 97 percent of the mass of the human body.

R. MEBANE

See also: ATOMS; CARBON; CHEMICAL BONDS; ELECTRONS AND POSITRONS; ELEMENTS; HALOGENS; IONS AND RADICALS; METALLOIDS; METALS; NEUTRONS; NOBLE GASES; PERIODIC TABLE; PROTONS.

Further reading:

Emsley, John. 2001. *Nature's Building Blocks: An A–Z Guide to the Elements.* New York: Oxford University Press.

ABUNDANCE OF MAJOR NONMETALS AND SELECTED METALS (MASS%)

Element	Universe	Earth's crust, oceans, and atmosphere	Human body
Oxygen	1.07	49.50	65.0
Carbon	0.46	0.08	18.0
Hydrogen	73.9	0.87	10.0
Nitrogen	–	0.03	3.0
Phosphous	–	0.12	1.0
Sulfur	0.04	0.06	0.26
Chlorine	–	0.19	0.15
Helium	24.8	–	–
Calcium	0.007	3.40	2.4
Magnesium	0.06	1.93	0.50
Potassium	–	2.40	0.34
Sodium	–	2.63	0.14
Iron	0.19	4.71	0.005
Zinc	–	0.013	0.003

Blank values (–) indicate less than 0.001 percent

NONSTOICHIOMETRIC COMPOUNDS

In a nonstoichiometric compound the numbers of constituent atoms cannot be expressed as whole numbers

Rust is a mixture of nonstoichiometric iron oxides, which form when iron gets wet. Rust crystals have defects that promote electrochemistry, which then creates more rust.

CONNECTIONS

- Nonstoichiometry is an important factor in the **CORROSION** of **IRON AND STEEL.**
- Some nonstoichiometric compounds, called zeolites, are important industrial **CATALYSTS** used in manufacturing **PETROCHEMICALS.**

In chemistry, stoichiometry refers to elements combining in chemical reactions. Derived from the Greek words *stoicheion,* "element" or "principle," and *metron,* "measure," stoichiometry refers to the relative number of atoms of different elements found in a chemical substance.

In general, when chemical substances react and the atoms of different elements combine to form compounds, they do so in whole-number ratios of their atoms. For example, when a mixture of hydrogen and oxygen burn in air, they form water, which has a simple ratio of constituent atoms:

$$2H_2 + O_2 \rightarrow 2H_2O$$

This principle of elements combining in repeatable, whole number ratios is such a tenet of chemistry that it is enshrined in chemical laws. The law of constant composition (expressed by French chemist Joseph-Louis Proust [1754–1826] in 1799) states that all pure samples of the same compound contain the same elements combined in the same proportions by mass. The law of multiple proportions (introduced by English scientist John Dalton (1766–1844) in 1803; see DALTON, JOHN) states that when the elements A and B combine to form more than one compound, the various masses of A that combine with a fixed mass of B are in a simple ratio.

The majority of chemical reactions follow these laws. The products of such reactions that contain two or more elements are referred to as stoichiometric compounds. Nonstoichiometric compounds are those in which the numbers of atoms of the elements present cannot be expressed as a ratio of small whole numbers. For example, the form of iron (II) oxide that forms at about 1830°F (1000°C) is called wüstite and does not conform to the stoichiometric formula FeO but has a composition in the range $Fe_{0.89}O$ to $Fe_{0.96}O$. Stoichiometric sodium chloride (NaCl) is readily converted to a nonstoichiometric form by heating it in sodium vapor, when it changes from colorless NaCl to yellow-brown $Na_{1.001}Cl$. Chemists often make nonstoichiometric compounds in this way, by subjecting the stoichiometric form to high temperatures and then quenching the product to room temperature.

Nonstoichiometric compounds are best known among the transition elements (those metals in the periodic table that have a partially filled d shell of electrons; see TRANSITION ELEMENTS). Examples include the hydrides, oxides, and sulfides of d-block elements (see PERIODIC TABLE) such as titanium (Ti), zirconium (Zr), vanadium (V), and niobium (Nb).

Many so-called insertion compounds are also nonstoichiometric. They consist of a stoichiometric host into which a neutral molecule or a metallic element has been inserted.

Nonstoichiometric compounds generally have a variable composition. Many are intensely colored and have metallic or semiconducting properties. Their chemical reactivity is different from that of the nearest equivalent stoichiometric compound.

SOME NONSTOICHIOMETRIC COMPOUNDS OF D BLOCK ELEMENTS

Compound	"x" within the range of	
titanium hydride (TiH_x)	1 to 2	
zirconium hydride (ZrH_x)	1.5 to 1.6	
niobium hydride (NbH_x)	0.64 to 1.0	
titanium oxide (TiO_x)	0.7 to 1.25 (rock salt lattice)	1.9 to 2.0 (alternate lattice)
vanadium oxide (VO_x)	0.9 to 1.20 (rock salt lattice)	1.8 to 2.0 (alternate lattice)
niobium oxide (NbO_x)	0.9 to 1.04 (rock salt lattice)	
zirconium sulfide (ZrS_x)	0.9 to 1.0	

CORE FACTS

- Nonstoichiometric compounds do not follow the laws of constant composition and multiple proportions.
- Many nonstoichiometric compounds are manufactured by raising stoichiometric compounds to high temperatures under specific conditions and then cooling them rapidly to room temperature.
- Nonstoichiometry arises from defects in the structure of solids that result in spaces unfilled by constituent atoms remaining empty or being filled by other substances.

The causes of nonstoichiometry

The possibility of nonstoichiometric compounds arises from the presence of defects in the lattice structures in crystalline solids. Many solids at normal temperatures and pressures consist of extended arrays of ions, with positively charged ions (cations) attracted to negatively charged ions (anions) in a lattice. In many cases, the ions are close packed, and the spaces between ions of the same type are either tetrahedral (four sided) or octahedral (eight sided) and are filled with other ions or atoms in a regular arrangement. For example, in sodium chloride, common salt (NaCl), each sodium ion occupies an octahedral space created by six surrounding chloride ions. In nonstoichiometric compounds the incomplete filling of these spaces gives rise to deviation from whole-number ion ratios (see CRYSTALS AND CRYSTALLOGRAPHY; IONS AND RADICALS).

Deviations from stoichiometry arise in three general ways. Some of the atoms or ions of one element may be missing from a conventional lattice arrangement, as in the case of wüstite (Fe_xO, where x is less than 1). Alternatively, some of the atoms or ions may be present in excess of requirements, as in the case of nonstoichiometric zinc oxide (Zn_xO, where x is greater than 1). A third possibility is that some atoms or ions are replaced by those of another kind, as in compounds of bismuth and tungsten (Bi_2Ti_x, where x can be greater or less than 3).

Distinguishing compounds

Nonstoichiometric compounds can be distinguished from stoichiometric forms because of their variable composition. They behave like solid solutions in that, like liquid solutions, the contribution of chemical constituents to the physical and chemical properties of the solution varies as the composition changes (see SOLUTIONS AND SOLUBILITY). For example, if a nonstoichiometric metal oxide is subjected to a range of partial pressures (effectively, concentrations) of oxygen, the composition of the oxide at equilibrium varies across a continuous range (see EQUILIBRIUM; PHASE TRANSITIONS).

If the same experiment is repeated for stoichiometric metal oxides, their composition should alter in a stepwise fashion at varying partial pressures of oxygen. In practice, if the reaction between oxygen and the metal or metal oxide is slow, a stoichiometric series of compounds may be difficult to distinguish from a nonstoichiometric series, as is the case for the stoichiometric series of molybdenum oxides (Mo_8O_{23}, Mo_9O_{26}, $Mo_{10}O_{29}$, $Mo_{11}O_{32}$, $Mo_{12}O_{35}$, and $Mo_{14}O_{41}$). The use of X-ray diffraction or electron microscopy techniques can distinguish gradual from abrupt changes in structure under different conditions and so can distinguish nonstoichiometric from stoichiometric series.

Classifying nonstoichiometric compounds

One approach to classifying nonstoichiometric compounds is to consider for binary (two-element) compounds which element is in excess of stoichiometric expectations and account for how this excess is brought about. Insertion compounds are classified in a different manner.

Binary compounds

A classic example of a nonstoichiometric binary compound with metal in excess is zinc oxide (ZnO). When crystals of stoichiometric zinc oxide (ZnO) are heated in zinc vapor at temperatures in the range 1110 to 2190°F (600 to 1200°C), the crystals convert to nonstoichiometric zinc oxide (Zn_xO, where x is slightly greater than one) and change color from white through yellow to red. When cooled to room

The tetrahedral molecules of silicate and aluminate are arranged into four- or six-membered rings, which form the subunits of a highly porous supercage structure called a zeolite. The high surface area of the zeolite molecules makes them ideal for use as catalysts, with many sites for the catalyzed reactants to bond.

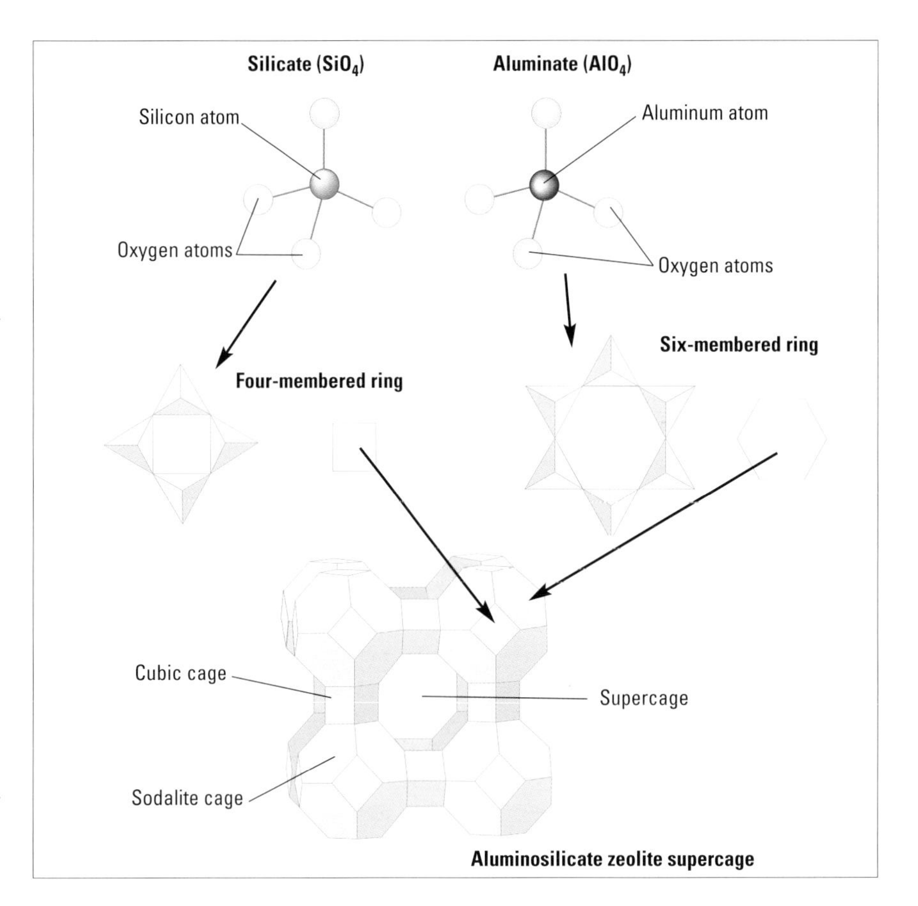

ALUMINOSILICATE ZEOLITES

Aluminosilicates are zeolites in which aluminum ions (Al^{3+}) replace some of the silicon ions (Si^{4+}) in a silicon and oxygen lattice. To maintain electrical neutrality, chemists usually top up the structure with sodium (Na^+) or hydrogen (H^+) ions. The building blocks of aluminosilicates are arrangements of silicon surrounded by four oxygen atoms or aluminum surrounded by four oxygen atoms, both of which have a tetrahedral shape. The tetrahedra combine to create four- or six-membered rings, which assemble to create forms such as sodalite and cubic cages enclosing open spaces called supercages (see the diagram above). The overall effect is to create a highly porous open structure with tunnels and chambers in which entering molecules can become trapped. Chemists can tailor the lattice to have pores of specific sizes to control the chemical traffic that enters and leaves. Aluminosilicates, which occur naturally as feldspars and clays (see CLAY), have a great many uses in manufacturing, both as industrial catalysts and potentially as compounds that adsorb (trap on their surface) pollutants.

A CLOSER LOOK

SEMICONDUCTORS

Within many nonstoichiometric compounds, cations of the same element in different oxidation states exist in the crystal lattice; this structure gives rise to semiconductive properties (see SEMICONDUCTORS). Where conductivity arises from excess of negative charge, as in the case of nonstoichiometric zinc oxide, the substance is called an n-type semiconductor. Where there is an excess of positive charge or a deficit of negative charge, as in the case of copper (I) oxide, the substance is called a p-type or positive-hole semiconductor. Combinations of n- and p-type semiconductors are the mainstay of the transistor industry, on which modern microprocessors—and computer technology—are based.

A CLOSER LOOK

temperature, the nonstoichiometric form has enhanced electrical conductivity compared with conventional zinc oxide. The red color and enhanced conductivity are attributed to the additional zinc atoms trapped within the lattice.

A wide range of tungsten (W) oxides exist, many of which are stoichiometric although they seem, at first examination, to be nonstoichiometric. For example, $WO_{2.9}$ is actually stoichiometric $W_{20}O_{58}$, and there are series of stoichiometric oxides with the general formula W_nO_{3n-2} that, as mixtures, can be mistaken for nonstoichiometric forms. However, true nonstoichiometric tungsten trioxide (WO_{3x}, where x is slightly less than 1) is obtainable by heating a stoichiometric tungsten trioxide crystal at about 2010°F (1100°C) under a controlled partial pressure of oxygen and allowing it to slowly cool. Gradual removal of oxygen from the trioxide makes it a semiconductor, and its color changes from pale yellow, through green to blue-green, and finally to black or dark brown. The resultant trioxides are examples of a nonstoichiometric binary compound deficient in its nonmetal component.

Cobaltous oxide (Co_xO, where x is slightly less than 1), where the metal component is deficient, does not have a stoichiometric equivalent (CoO). Whichever way cobaltous oxide is manufactured, nonstoichiometric forms result. Electrical neutrality of the lattice is maintained by any missing Co^{2+} ions being compensated for by two Co^{3+} ions elsewhere in the structure.

Uranium dioxide (UO_{2x}, where x is greater than 1) is an example of a compound where the nonmetal is in excess. The broad range of values of x between 1.0 and 1.125 is attributed to there being two lattice arrangements: additional oxygen atoms in forms of UO_2, where x is less than 1.0625, and oxygen vacancies in U_4O_9, where x is greater than 1.0625.

Titanium oxide (TiO_x, where x has values in the range 0.85 to 1.18) is unusual because of its wide range of composition and the metallic character of its nonstoichiometric forms. Titanium oxide, like rock salt, has a cubic structure. Stoichiometric titanium oxide has an unusually large number of vacant spaces in its lattice (about 15 percent). By altering oxygen partial pressures when the stoichiometric form is heated or by adding excess titanium, a wide range of nonstoichiometric forms are possible. The metallic character of these forms is attributed to delocalization of overlapping d orbital electrons.

Nonstoichiometric insertion compounds

Tungsten bronzes are based on tungsten trioxide (WO_3), in which a wide range of metals and some nonmetals can be inserted. Sodium tungsten bronzes have the formula Na_xWO_3, where x ranges widely, between less than 0.25 to as high as 0.98. These compounds are created in several ways, including electrolyzing or heating mixtures of stoichiometric tungsten compounds. Sodium tungsten bronzes range in color from blue, where x is 0.4, to yellow-gold at 0.98. Where x is less than 0.25, the bronzes are semiconductors; where x is greater than 0.25, they are metallic.

Intercalation compounds are host compounds that accommodate other molecules ("guests") in holes in the lattice or between layers in the lattice. Clathrates, for example, are host molecules that hold guest molecules within isolated cavities within their structure. They include methane hydrates—also called methane ice—where crystalline structures form when methane and water combine under the high pressure and cool temperatures found below about 1,000 feet (300 m) on the ocean floor. Methane hydrates trap other molecules, including other hydrocarbon gases such as ethane and propane (see HYDROCARBON). Perhaps 10 percent of the ocean floor harbors methane hydrates and thus, they represent a huge potential source of fossil fuel if they can be exploited safely (see NATURAL GAS).

Zeolites, another group of intercalation compounds, are molecular sieves in which guest molecules can move through a network of tunnels in the lattice. Several other categories of intercalation compounds exist, amounting to many hundred different kinds of host molecules with each capable of harboring a wide range of guest molecules. The creation and use of intercalation compounds is a burgeoning field of modern chemistry.

T. DAY

See also: CLAY; CRYSTALS AND CRYSTALLOGRAPHY; DALTON, JOHN; EQUILIBRIUM; HYDROCARBONS; IONS AND RADICALS; MINERALS AND MINERALOGY; NATURAL GAS; PERIODIC TABLE; SEMICONDUCTORS; SOLUTIONS AND SOLUBILITY; TRANSITION ELEMENTS.

Further Reading

Brown, T. E., H. E. LeMay, B. E. Bursten, and J. R. Burdge. 2002. *Chemistry: The Central Science*. Upper Saddle River, N.J.: Prentice-Hall.

Gusev, A. I., et al. 2001. *Disorder and Order in Strongly Nonstoichiometric Compounds: Transition Metal Carbides, Nitrides, Oxides*. Berlin: Springer-Verlag.

Housecroft, C. E., and A. G. Sharpe. 2001. *Inorganic Chemistry*. Upper Saddle River, N.J.: Prentice-Hall.

NORTH AMERICA

North America is a continent situated on the North American plate

North America is located in the Northern Hemisphere and is bounded on the west by the Pacific Ocean, on the east by the Atlantic Ocean, on the north by the Arctic Ocean, and on the south by the Caribbean Sea. This continent includes one-sixth of the world's dry land area. Roughly triangular in shape, it is the third largest continent, covering 9,362,000 square miles (24,341,000 km^2) of land. Only Asia and Africa cover larger areas, and only Asia has more than North America's 37,000 miles (59,500 km) of coastline.

North America is defined in different ways according to different fields of study. If described in terms of political boundaries, it includes Canada, Greenland, the United States, Mexico, Belize, Costa Rica, El Salvador, Guatemala, Honduras, Nicaragua, and Panama. If described as geographical elements, it includes the northern continent in the Western Hemisphere as well as the Arctic Archipelago, the Bahamas, the Greater and Lesser Antilles, the Queen Charlotte Islands, and the Aleutian Islands.

North America is a land of extremes. In the north, it stretches 5,400 miles (8,700 km) across from Alaska's Aleutian Islands to Newfoundland. In the south, its narrowest part—between the Pacific Ocean and the Caribbean Sea—is only 31 miles (50 km) wide. From north to south, North America stretches more than 5,000 miles (8,000 km) at its longest expanse. The highest point in North America is 20,320 feet (6,194 m) above sea level at the top of Mount McKinley in Alaska. The lowest point is 282 feet (86 m) below sea level in Death Valley, California. In its far northern reaches, the soil is permanently frozen, while in the southernmost parts the climate is tropical.

The North American plate

The continent of North America is located on the North American plate, which is bounded on the east by the Eurasian plate and the African plate; on the south by the Caribbean, the Cocos, and the South American plates; and on the west by the Pacific plate. The North American plate is growing slowly as it moves from east to west. New seafloor is forming along the trailing edge of the plate at the spreading axis in the central Atlantic Ocean. The eastern continental margin is blanketed by thick layers of marine sediment over the thinned continental rock (see CONTINENTAL SLOPES). The spreading rate in the Atlantic is about 0.3 inches (1 cm) per year. This type of margin is called a diverging plate boundary. On the western edge of the North American continent, a converging margin occurs, as one plate slides beneath the other and is melted. Because the edge of North America is made up of lightweight continental rock, it rises above the edge of the Pacific plate. In addition, large sections of the plate boundary simply slide past the neighboring plate, forming transform plate margins (see PLATE TECTONICS).

North America was the first continent to approximate today's size and shape (about 600 million years ago), and it contains some of the oldest rocks in the world. The continent is built around a stable mass of rock called the Canadian Shield. The shield, located in eastern Canada, contains the remains of at least four ancient continents that were fused together 2 billion years ago. The ancient Appalachian Mountains grew to the southeast. To the west grew a younger mountain belt called the North American Cordillera, which starts in the Aleutian Islands and stretches southeast through Central America to the Andes. The cordillera predominates from Alaska to Central America and widens into the Rocky Mountains in the central portion of the continent (see MOUNTAINS).

This true-color satellite picture, showing North and Central America, was taken from 80 degrees west on an equatorial orbit at an altitude of 22,370 miles (36,000 km). The light brown areas to center left are the desert regions of Arizona and Utah, and to the right of these are the Rocky Mountains.

CORE FACTS

- North America is a continent located in the northern half of the western hemisphere on the North American plate.
- It is the third largest continent today and reached something like its modern size and shape roughly 600 million years ago. It includes some of the oldest rocks in the world.
- The four major climate zones in North America are the arctic, cool temperate, warm temperate, and tropical humid zones.

CONNECTIONS

- The **CONTINENT** of North America is bordered by the **ATLANTIC OCEAN** on the east and the **PACIFIC OCEAN** on the west.
- The eastern edge of the North American **TECTONIC** plate forms the Mid-Atlantic Ridge.

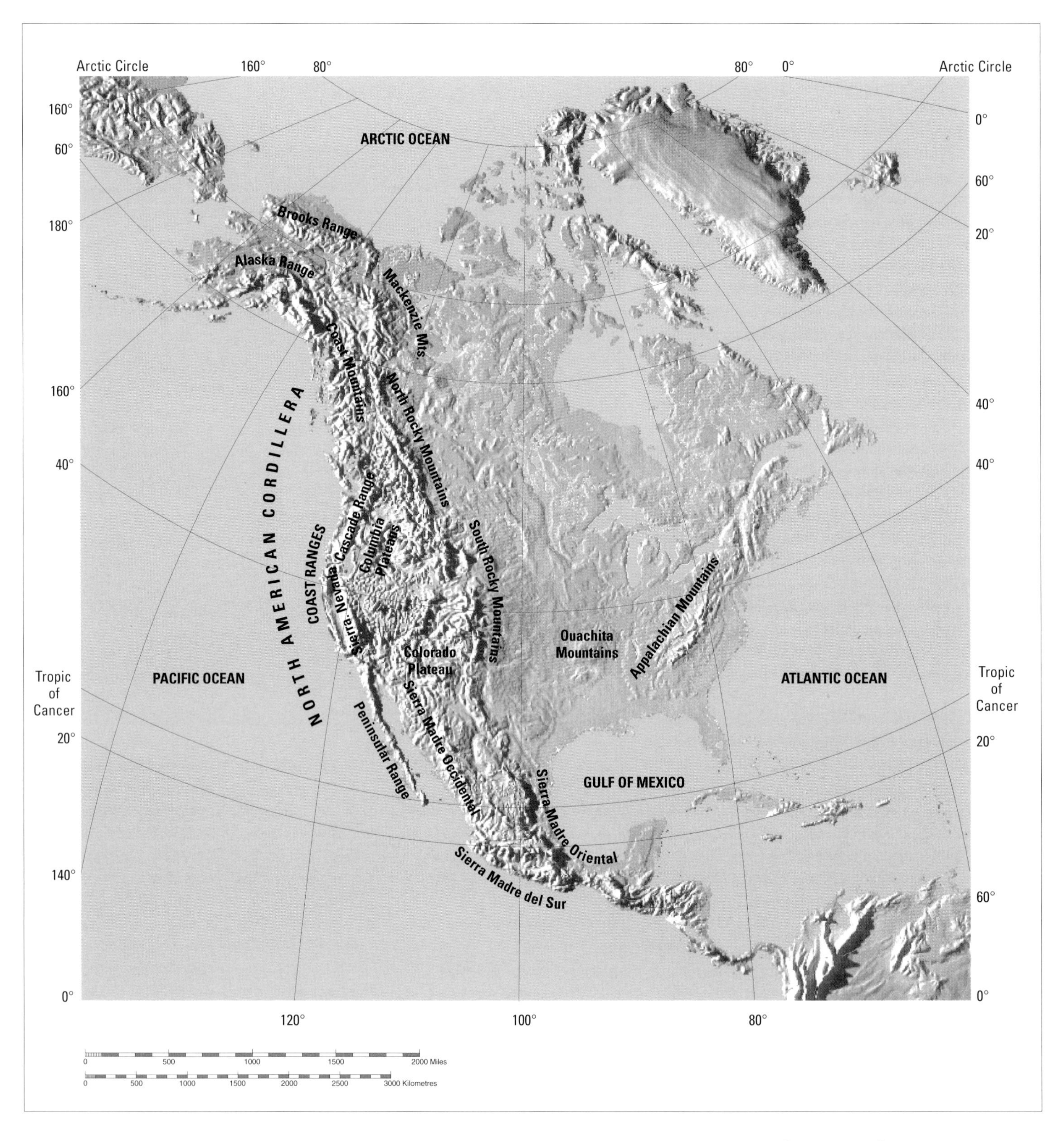

The diagram above shows the distribution of mountains in North America. The North American Cordillera in the west represents the tallest and youngest mountain belt, stretching from Alaska through Central America.

Weather systems

Although most of North America falls within the temperate zone, the northern portion of the continent is very cold, while a small portion to the extreme south is tropical (see TEMPERATE REGIONS; TROPICAL REGIONS). The large northern landmass and frozen winter seas make winters long and cold as far south as the Ozarks. Much of the far northern land has permanently frozen subsoil and has temperatures 6 to 8°F (3 to 4°C) cooler than the average for their latitude (see PERMAFROST). Because the North Pacific is warmed by the Japan current, the land of the Pacific northwestern section of the United States and Canada is 8 to 10°F (4 to 5.5°C) warmer than the average for its latitude (see CURRENTS, OCEANIC).

The continent's climate shows great contrast between the coastal and inner continental areas. Warm temperatures extend northward along the west coast to Alaska, while a great cold area extends south along the Mackenzie River valley and the Canadian Shield over the center of the continent.

The majority of North America's land receives adequate rain for crop and livestock production, although the interiors can be very dry. The western mountains of Canada and the United States have wet slopes facing west and dry eastern slopes facing the interior. The jet stream, a strong airflow moving over the middle of the continent, drives most storms from west to east (see JET STREAM).

Of the major air masses that move over North America, the most influential are the polar continental, the tropical gulf, and the Pacific polar winds. These air masses follow somewhat predictable patterns and determine the overall climate of North America, as well as the weather at particular times of the year. The polar continental is responsible for the cooler-than-average temperatures, for the latitude, that prevail over much of the continent. This cool-to-cold high pressure dome over the Canadian Shield sends winds sweeping over Labrador and New England and southward across the Great Lakes and the Great Plains in winter (see AIR PRESSURE AND BAROMETERS). Tropical gulf air drives the polar continental air mass north in the summer. The tropical Gulf's low-pressure systems develop in the Mississippi River basin as the center of the continent warms up. Reaching its height in July, this warm, moist, and unstable air mass forms two loops to the northwest and northeast and brings warm temperatures into Alberta and Quebec. The polar Pacific air mass operates on the west coast from northern California to Alaska, where it brings warmer, moist air to these areas, especially in winter.

Climate zones

There are four major climate zones in North America: the Arctic, the cool temperate, the warm temperate, and the tropical humid zones. The great extremes of temperature, precipitation, and variations in the length of growing season in North America result from the wide continental expanse from east to west and the north-south reach of the continent from the Pole to near the equator.

The Arctic zone includes the northern parts of the Canadian Shield and Alaska, the Canadian Arctic archipelago, and Greenland. Temperatures below 0°F (−18°C) last for five to seven months, with frost for eight to ten months out of the year. The frost-free season is less than 60 days long.

The cool temperate zone includes Newfoundland to Alaska and from Hudson Bay to the Ohio River. The winters, which begin after October, are long and harsh: temperatures fall and stay low until April or early May. North of the Great Lakes, temperatures below 0°F (−18°C) are common in January and February, and temperatures as low as −20 to −80°F (−29 to −62°C) have been recorded. After the transition to spring, the growing season has between 90 and 120 frost-free days. In the summer, moist Gulf air masses bring sufficient rain for crops.

The warm temperate zone stretches from the southeastern coast of the United States to the Mississippi River and along the Gulf coast (see GULF OF MEXICO). There, the tropical gulf air mass extends the frost-free season beyond 200 days. Polar continental air comes south in November, but winters are mild, with average February temperatures of between 40 and 54°F (4 to 12°C). Rainfall may

ICELAND IS BEING TORN APART

Iceland is a volcanic island sitting on the Mid-Atlantic Ridge, which divides the North American plate from the Eurasian plate. Normally, these plate boundaries are submerged in ocean basins, but Iceland is one of only a few places where a spreading zone rises above the sea and hence, it is a perfect place for geologists to observe a spreading zone in action.

Although the average movement along the Mid-Atlantic Ridge is several inches per year, volcanic events in Iceland may result in movement of 3 to 13 feet (1 to 4 m) in a matter of days, after which years may go by with no movement at all. In these ridges, called constructive plate boundaries or margins, new volcanic material is rising up, adding to the edges of each plate and pushing outward. Iceland is growing as it is being torn apart.

A CLOSER LOOK

Greenland is at the extreme northern limit of North America. It experiences an arctic climate, and much of the mountainous land surface has a permanent covering of snow.

THE GRAND CANYON

The Grand Canyon reveals the North American continent's history. The slow uplift of the Colorado Plateau coupled with the steady flow and erosion of the Colorado River produced the Grand Canyon in west-central North America. The slow uprising of the plateau rocks over 5 million years, the continuous presence of the Colorado River system to erode the land, and the arid climate that provided little vegetation to resist erosion all contributed to this unique natural wonder.

Geologists can study rock-layer records up to 14,800 feet (4,500 m) deep in the canyon. At the bottom of the Grand Canyon, for example, the top layer of the continent's Precambrian (more than 570 million years ago; see PRECAMBRIAN) foundation is revealed. Although there are gaps (called unconformities) in the record owing to layers that have been completely eroded, a detailed record of the continent's Precambrian and Paleozoic history is revealed in the canyon.

A CLOSER LOOK

The San Andreas Fault marks the meeting point of the North American plate and the Pacific plate. Relative movement of the two plates has created dramatic landforms, such as this fault scarp parallel to the highway north of Los Angeles.

be heavy in summer, hurricanes a hazard in the southeastern states, in the lower Mississippi Valley, and along the Gulf of Mexico (see GULF OF MEXICO; HURRICANES).

To the west of the warm temperate zone and into the Great Plains and the Rocky Mountains, there is a mix of highland, steppe, and desert climates. The highland climate of the Rockies extends west into the Sierra Nevada and, farther south, continues throughout the mountains of the Western Sierra Madre in Mexico.

SAN ANDREAS FAULT: WAITING FOR THE BIG ONE

The San Andreas Fault is a transform boundary between two plates—the North American plate and the Pacific plate (see FAULTS). As the Pacific plate slides to the northwest along the North American plate, deep movement of the plate results in movement on the surface called earthquakes (see EARTHQUAKES). The longer the surface movement is delayed beyond the subsurface movement of the sides, the greater the buildup of stress along the fault. Hence the phrase "waiting for the big one" describes the inevitable large surface correction that must eventually result from years of steady subsurface movement of the plates. This last occurred in 1906, when 28,000 buildings in San Francisco were destroyed in an earthquake.

A CLOSER LOOK

Beyond the Rockies in Oregon and much of California, a variation of the temperate climate exists. There summers are dry because tropical continental air dominates, and frosts occur only when polar continental air pushes to the coast in winter. Precipitation is insufficient for growing crops during the summer, and irrigation is used extensively.

A tropical humid climate prevails in Central America, where there is no winter, and even the coldest month averages 64°F (18°C) or warmer. Rainfall varies from 45 to 80 inches (114 to 203 cm), as the easterly trade winds bring moisture onshore. Summer hurricanes are a constant threat in this area and cause much damage to buildings and crops.

P. WEIS-TAYLOR

See also: CONTINENTAL SLOPES; CURRENTS, OCEANIC; EARTHQUAKES; FAULTS; GULF OF MEXICO; HURRICANES; JET STREAMS; MOUNTAINS; PERMAFROST; PRECAMBRIAN; TEMPERATE REGIONS; TROPICAL REGIONS.

Further reading:

Bradshaw, M., G. W. White, and J. P. Dymond. 2004. *World Regional Geography: Global Connections, Local Places.* Boston: McGraw-Hill.

Levin, H. L. 2003. *The Earth through Time.* 7th ed. Fort Worth: Harcourt, Brace, Jovanovich.

NORTH SEA

The North Sea, bordered by Britain and mainland Europe, is an extension of the Atlantic Ocean

The North Sea is an extension of the Atlantic Ocean. It lies between Britain and mainland Europe, bordered in the east by Norway and Denmark and in the south by Germany, Holland, Belgium, and France. The English Channel connects the southern North Sea to the Atlantic, and in the north, between Norway and Scotland, it merges into the North Atlantic and Norwegian Sea. A broad channel between Norway and Denmark, the Skagerrak, leads to the Baltic Sea.

The North Sea covers an area of about 220,000 square miles (575,000 km^2). Its maximum length is 600 miles (950 km) and the maximum width 400 miles (650 km). It is relatively shallow, with an average depth of 308 feet (94 m). The seafloor is irregular, and there are some deeper areas, notably along the Norwegian coast, where the depth can be more than 2,140 feet (650 m). The maximum recorded depth is about 2,400 feet (730 m). One area, the Dogger Bank, is particularly shallow, with a depth as little as 50 feet (15 m). This sandy bank is some 160 miles (260 km) long and 60 miles (100 km) wide and occupies a central position between Britain and Denmark.

There are a number of regular currents in the North Sea. They are caused mainly by the North Atlantic Current, which enters the North Sea from the Norwegian Sea in the north and, from the south, via the English Channel. In addition, the Baltic Current, originating in the Baltic Sea, enters the North Sea from the east, along the Skagerrak (see CURRENTS, OCEANIC). Various rivers, including the Thames, the Elbe, and the Rhine, flow into the sea.

Some 2.5 million years ago, in the Pliocene epoch, the area now occupied by the North Sea was a large plain of dry land (see CONTINENTAL SHELVES). This plain was flooded by rising sea levels, which finally separated Britain from France around 9000 to 8000 B.C.E. By about 1000 B.C.E., the coastlines seen today became recognizable.

Important resources of the North Sea

The North Sea yields some 5 percent of the entire world's catch of fish—mainly cod, haddock, herring, and flatfish such as plaice and sole. Many fish use the shallow waters as a spawning and feeding ground, and numbers of several species have been very high.

Overfishing has become more and more of a problem in recent years, and quotas have had to be established to prolong the survival of the fishing industry. Herring, once a staple fish, has suffered a steep decline, and stocks are now particularly low. The shallow waters of the Dogger Bank often hold large quantities of several species, including cod, haddock, herring, mackerel, and plaice. Cod fishing has a distinctly seasonal aspect. Off the Norwegian coast, for example, cod is harvested as it migrates to its spawning grounds in winter.

The satellite image above shows a large fog bank (pale yellow) in the North Sea between the British Isles and the coasts of northern Europe and Scandinavia.

The North Sea is one of the world's key non-OPEC (Organization of the Petroleum Exporting Countries) oil and gas producing regions. It contains Europe's largest oil and natural gas reserves, which were first discovered in the North Sea in the 1960s. Semipermanent platforms on stiltlike legs are used as supports for the drilling rigs, and clusters of these now reside in certain areas, from the central North Sea to the waters northeast of the Shetland Islands. Offshore oil production reached a peak in 1999 but has declined in recent years. In contrast, natural gas production has continued to increase year to year.

M. WALTERS

See also: ATLANTIC OCEAN; CONTINENTAL SHELVES; CURRENTS, OCEANIC.

Further reading:

Gibbs, W., and J. Rajesh. 2003. *North Sea Saga: The Oil Age in Norway*. Oslo, Norway: Horn Forlag.

Glennie, K. W., ed. 1998. *Petroleum Geology of the North Sea: Basic Concepts and Recent Advances.* Malden, Mass.: Blackwell Science.

CONNECTIONS

- **MARINE EXPLORATION** of the North Sea led to the discovery of **OIL** and **NATURAL GAS**.
- **POLLUTION** of the North Sea is a problem caused by industry around its shores.

NOVAS AND SUPERNOVAS

Novas and supernovas are stars that suddenly explode and temporarily shine many times brighter than before

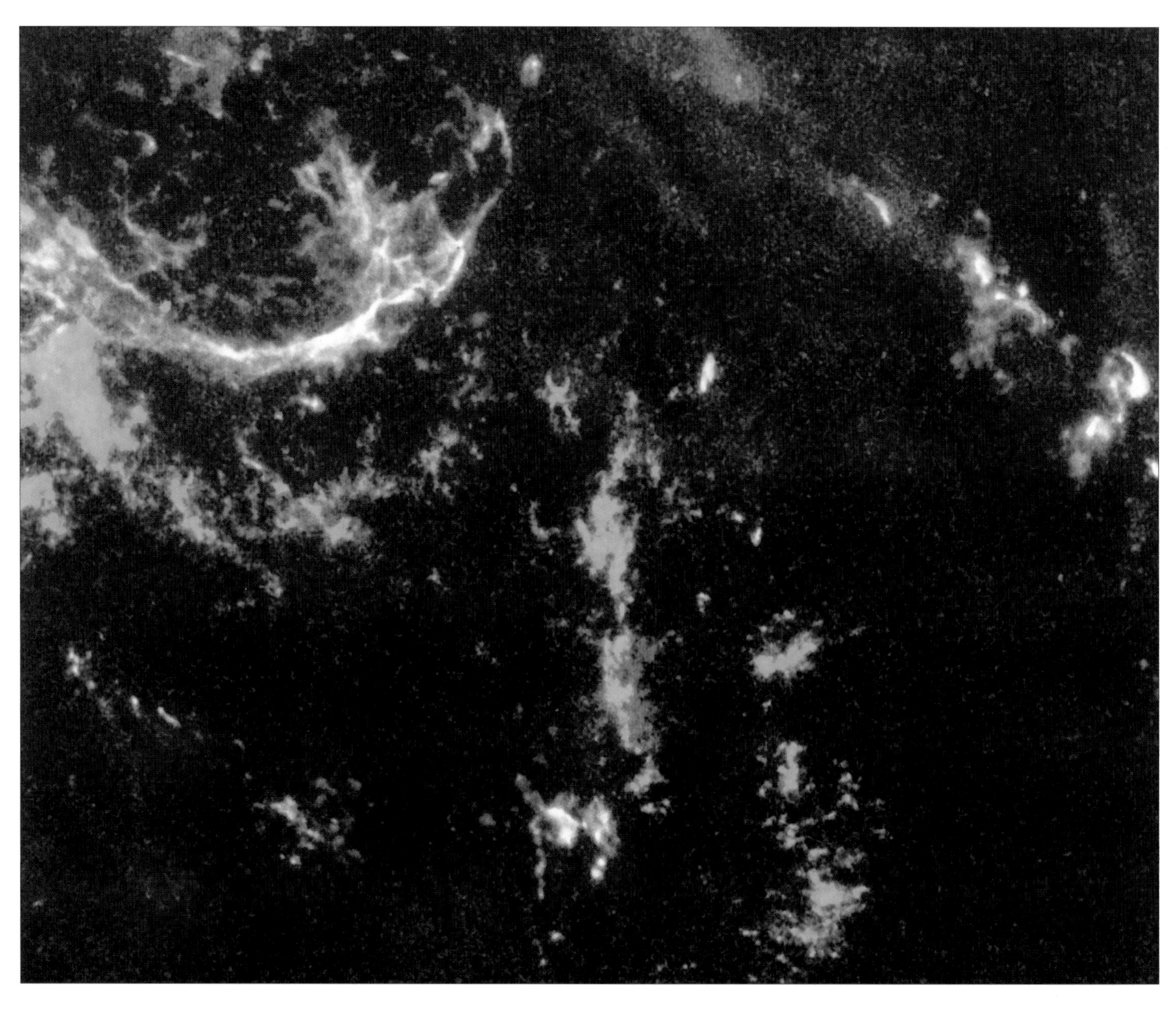

The remnant of a supernova photographed by NASA's Hubble Space telescope.

Novas and supernovas are stars that suddenly explode and flare up and shine many times brighter than before. The word *nova* comes from the Latin word for "new": oftentimes a nova is so bright compared with its previous luminosity that it seems as if a new star has formed. As its name suggests, a supernova is even more striking in brightness. Some supernovas are so bright that they can be seen with the unaided eye from Earth, even though they are in other galaxies. Supernovas are named sequentially for the year in which they were detected—thus, the first to be detected in 1987 was called Supernova 1987A, the second was 1987B, and so on.

Novas

Novas are common, with astronomers logging dozens or more each year. Astrophysicists believe that almost all novas occur in close binary systems—pairs of stars that orbit one another (see BINARY STARS). One star of the pair is originally much more massive than the other. Every star undergoes an evolutionary process at a speed that depends on its mass. The more massive star in the binary system evolves more rapidly

CONNECTIONS

- Some **PULSARS** originate in the aftermath of a supernova explosion.

CORE FACTS

- Novas and supernovas are exploded stars that temporarily shine thousands or millions of times brighter than before.
- Novas are more common than supernovas: only six supernovas are known to have occurred in the Milky Way galaxy during the past 1,000 years.
- Supernovas can be divided into Type I and Type II supernovas, depending upon the mechanisms of the explosion.

than its partner. The more massive star therefore consumes the hydrogen in its core more quickly, fusing the hydrogen into helium and then into carbon and oxygen (see FUSION; NUCLEOSYNTHESIS).

The larger star balloons outward to become what is called a red giant (see RED GIANTS). As it does so, it may envelop its smaller, companion star. Over time much of the larger star's mass is lost, and it collapses into a white dwarf—an enormously dense object with a mass similar to that of the Sun but with a diameter similar to that of Earth (see WHITE DWARFS). In a white dwarf, atoms are packed as tightly together as the laws of physics allow without combining to form a uniform mass of neutrons or collapsing to become a black hole (see BLACK HOLES; NEUTRON STARS).

Meanwhile, the two stars in the binary system orbit ever closer to one another. Owing to the density of the white dwarf, the gravitational field strength becomes so strong that the white dwarf draws hydrogen off the companion star. This hydrogen does not all fall directly onto the star but instead forms a flat accretion disk in space around the white dwarf. Gradually the hydrogen at the surface of the white dwarf comes under greater and greater pressure and heat until it bursts into fusion, the result being the enormous flash of light that characterizes a nova.

A nova explosion increases a star's brightness by a factor of several thousands. Most novas follow a characteristic pattern of remaining bright for a short time—a few hours or days at the most—and then gradually dimming over a year or more. The pattern of how the light changes with time is known as a light curve. Unfortunately, novas and supernovas have very similar light curves, so it is impossible to distinguish between the two on that evidence alone.

It was once thought that, after the explosion, the flow of hydrogen-rich material resumes almost immediately and the whole process that produced the initial outburst repeats itself, resulting in another explosion 1,000 to 10,000 years later. However, more recent research has shown that the cycle may take place at much longer intervals—every 100,000 years or more, depending on the mass of the white dwarf and the distance between the binary pair. Scientists believe that the aftershock of the nova eruption separates the members of the binary system and slows the flow of material until, after a long time, the two stars move close together again.

SUPERNOVAS IN HISTORY

Because ancient peoples frequently looked to the skies, they often saw supernovas that were visible to the unaided eye. Old astronomical records continue to be an important source of information for astronomers who are seeking to understand what causes supernovas. However, some records are strangely incomplete: for example, astronomical records in Europe make no mention of the supernova of 1054, which surely would have been visible and which was recorded by astronomers in other parts of the world.

The supernova of 1006, the brightest in recorded history, was observed by astronomers in China, Japan, Korea, Europe, and the Middle East. By comparing different observations, modern-day astronomers think they have found the supernova remnant produced by the blast—a sphere of gas that seems to be about the right size to have been expanding for about 1,000 years.

HISTORY OF SCIENCE

Supernovas

Supernovas are far more energetic and dramatic than novas and can shine with a brightness 10 billion times greater than Earth's own Sun. Only six supernovas are known to have happened in the Milky Way galaxy during the past 1,000 years. They occurred in 1006, 1054, 1181, 1572, 1604, and at one uncertain date during the mid 17th century.

One supernova in the constellation of Taurus (see CONSTELLATIONS) in the year 1054 was so bright that it could be seen during daytime for 23 days and at night for almost two years. This supernova created the Crab Nebula (see NEBULAS).

It is impossible to know when a supernova will occur, but observations of supernovas in other galaxies suggest that the Milky Way should have between one and several supernovas every century. The fact that none has been observed in almost 400 years could be nothing more than a statistical fluctuation; more likely, supernovas have been occurring but have been blocked from our view by interstellar dust.

The massive explosion of a supernova ejects huge amounts of gas from the star millions of miles into space. Over time, the result is a remnant many light-years in diameter. (One light-year is 5.88 trillion mi., or 9.46 trillion km.) The energy released by the explosion hurls the gas in all directions at very high speed, and the gas rams into the ordinary interstellar gas and dust that is sprinkled throughout a

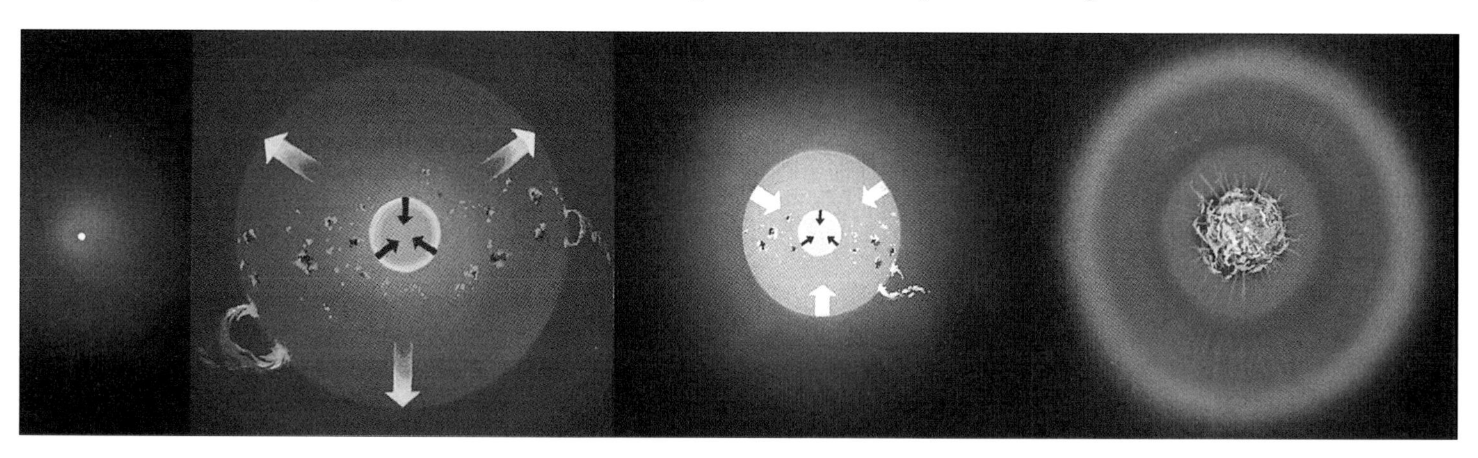

This illustration shows the stages leading to a supernova explosion. Stage 1 (left) shows a normal star powered by the fusion of hydrogen to form helium. When the hydrogen is depleted, the helium ignites and causes the star to expand (stage 2). Heavier elements burn in shells around the core. Eventually, the core collapses (stage 3). The core rebounds, and the shock tears the star apart (stage 4).

SUPERNOVA 1987A

The eruption of Supernova 1987A in the Large Magellanic Cloud, the nearest neighboring galaxy to the Milky Way, was a stroke of remarkably good luck for astronomers. Because the star that erupted had been photographed before the explosion, scientists had an unprecedented before-and-after look at the effects of a supernova. Experiments in Japan and Michigan registered neutrinos from the blast. (Neutrinos are a type of elementary particle; see NEUTRINOS). Shortly after the supernova was detected, astronomers around the world were training their most advanced instruments on the object. They have used the observations of the supernova remnant—which is all that can be seen right now—to understand the aftermath of this supernova explosion and thus the mechanism of the explosion itself. As the supernova remnant unfolds over the decades to come, astronomers on Earth will be watching closely.

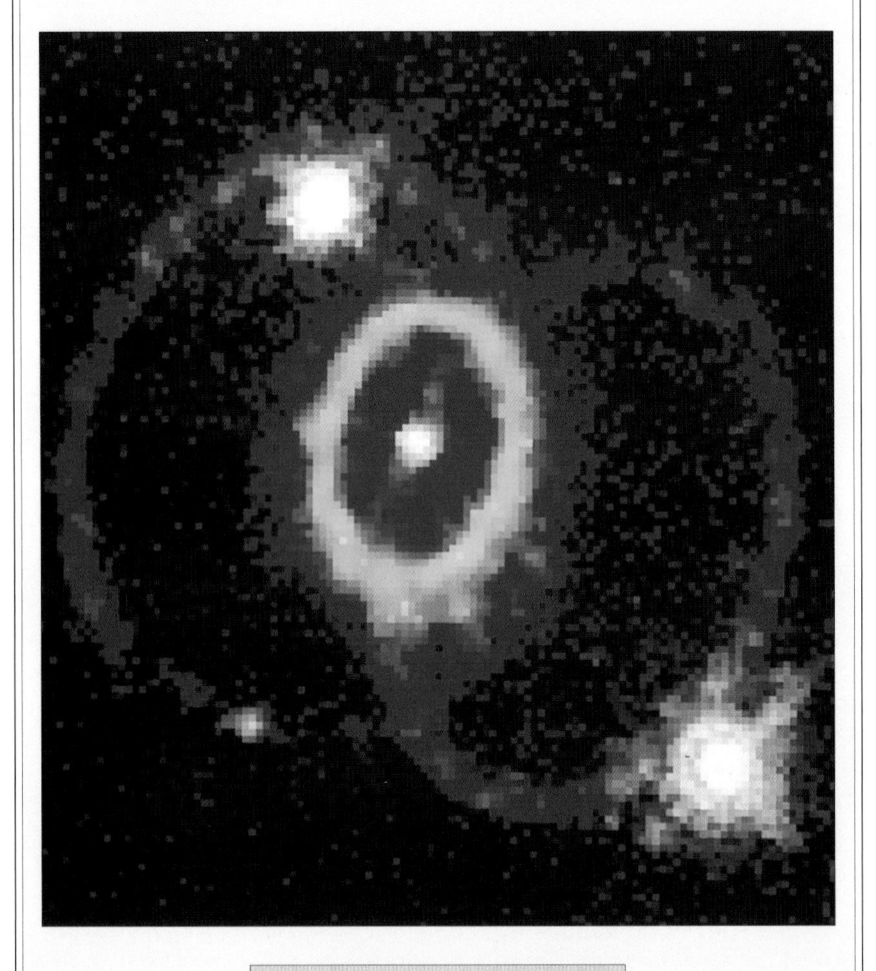

A CLOSER LOOK

galaxy. The force of the impact and the resulting glow of the interstellar dust and gas produces a nebula. The mass of gas thrown into space by the supernova is known as a supernova remnant.

Two types of supernova explosion

The term *supernova* was proposed by Swiss astronomer Fritz Zwicky (1898–1974) and German astronomer Walter Baade (1893–1960) in 1933; they suggested that a supernova was more than simply a bright nova and instead was something very different, caused by a star being blown to pieces. Astronomers place supernovas into two categories—Type I and Type II. The two types can be distinguished by the presence or absence of hydrogen in their spectra (see FRAUNHOFER LINES; SPECTROSCOPY). Their light curves also differ. Type I supernovas, which have hydrogen in their spectra, stay at their brightest for a very short time and gradually become dimmer; Type II supernovas, which lack hydrogen in their spectra, remain at their peak brightness for a longer period of time but then lose their brightness more rapidly.

The two types reflect differences in the stars that produce the supernovas. Astronomers group stars into two populations. Population I stars (including the Sun) are believed to be younger and to contain heavier elements made in earlier stars. Population II stars are older and consist mainly of hydrogen and helium. Typically, Population I stars are found in the central disks and arms of spiral galaxies, while Population II stars are found in globular clusters and in halos surrounding galaxies (see ASTROPHYSICS; GALAXIES). As the nature of supernovas became better understood, it turned out that Type I supernovas are generally produced by Population II stars (old), and Type II supernovas are generated by Population I stars (young).

Astrophysicists believe that Type II supernovas occur when a red giant, a large, massive star, runs out of fuel. When all the hydrogen in the core has been fused to helium, the helium fuses to form carbon and oxygen, and then carbon is fused to form manganese and neon. Eventually, the star's core is fusing silicon to produce iron. However, iron atoms cannot be fused together, so at that point the fusion reaction stops, and the star's core stops producing energy. As a result, the core collapses in on itself, and a huge shock wave ripples out through the star and throws its atmosphere into space: a supernova. All that is left of the red giant is a superdense neutron star (see NEUTRON STARS).

Scientists are less certain about the mechanism in Type I supernovas. One possibility is similar to that in a nova, with a binary system made up of a white dwarf drawing in material from another star. If the white dwarf could attract enough material, it might become massive enough to collapse in on itself to form a neutron stars, produce a supernova in the process.

Astronomers have suggested that the solar system was formed following a supernova that occurred about five billion years ago. According to this theory, the heavy elements in today's solar system were created inside supernovas before forming a vast cloud. The shock wave from another supernova caused the cloud to collapse in on itself and form the solar system.

V. KIERNAN

See also: ASTROPHYSICS; BINARY STARS; BLACK HOLES; CONSTELLATIONS; FRAUNHOFER LINES; FUSION; GALAXIES; NEBULAS; NEUTRINOS; NEUTRON STARS; NUCLEOSYNTHESIS; QUASARS; RED GIANTS; SPECTROSCOPY; STARS; UNIVERSE; WHITE DWARFS.

Further reading:

Höflich, P., P. Kumar, and J. C. Wheeler, eds. 2004. *Cosmic Explosions in Three Dimensions: Asymmetries in Supernovae and Gamma-ray Bursts.* New York: Cambridge University Press.

NUCLEAR PHYSICS

Nuclear physics is the branch of physics that studies the atomic nucleus and the particles inside

The Large Electron-Positron collider in Geneva, Switzerland, helps nuclear physicists in their search for new subatomic particles. The underground path of the collider can be seen as a large circle. (The smaller circle marks the path of a proton-antiproton collider.)

Nuclear physics is the study of the atomic nucleus. After British physicist J. J. Thomson (1856–1940) discovered electrons in 1897 (see ELECTRONS AND POSITRONS), he suggested that atoms have a structure similar to a plum pudding, in which the electrons are the plums. Then, in 1911, one of Thomson's former students, New Zealand physicist Ernest Rutherford (1871–1937), found that atoms are not puddinglike at all. They are mostly empty space, with the electrons surrounding a small, dense mass—the nucleus—at the center.

In 1898 French physicist and chemist Marie Curie (1867–1934) had discovered the element radium and showed that it gives off positively charged rays, which Rutherford named alpha particles (α-particles). Electrons carry negative charges, so the interior of an atom must contain positive charges to balance them. With this idea in mind, Rutherford decided to use α-particles to investigate the structure of atoms by firing a stream of them through a small piece of thin gold foil. To his surprise, some of the α-particles were "bounced" back—as he later said: "It was almost as incredible as if you fired a fifteen-inch shell at a piece of tissue paper and it came back and hit you."

Rutherford proposed that the atom has a small, heavy nucleus made up of positive particles similar to the α-particles, which they were repelling. He named these protons and discovered that the nucleus of hydrogen is made up of just one proton (see PROTONS). Heavier elements, he concluded, have an increasing number of protons in the nucleus. In 1917 Rutherford made a further surprising discovery. If nitrogen was bombarded with α-particles, it gave off

CORE FACTS

- The nucleus, or dense mass at the center, of an atom comprises protons, which are positively charged, and neutrons, which are neutral. Both are called nucleons.
- Nearly all the mass of an atom is concentrated in the nucleus; nucleons are more than 1,836 times heavier than electrons.
- Nucleons are affected by three forces: the strong nuclear force binds neighboring nucleons together in the nucleus; the electric force causes protons to be repelled by one another; and the weak nuclear force can change neutrons into protons, and vice versa.

CONNECTIONS

- **PROTONS** and **NEUTRONS** in the nucleus occupy **ENERGY** levels much like **ELECTRONS** in the outer regions of the **ATOM**.

GLUONS

Besides the photon, physicists have discussed the existence of other "exchange particles" involved in nuclear forces. For example, there is a very weak gravitational force that exists between subatomic particles. The gravitational force is transmitted by the graviton.

What prevents quarks and whole nucleons from drifting off on their own is the strong nuclear force, and the exchange particles involved in this interaction have been named gluons. Gluons perform the same job in controlling relations among nucleons that photons do in exchanging electromagnetic energy between electrons. However, gluons have several peculiar properties. The farther electrons are separated, the fewer photons they exchange. The opposite is the case with gluons. The closer quarks are, the fewer gluons pass between them; however, as the distance between quarks increases, the rate of gluon exchange increases also. So gluons prevent quarks from breaking free; in fact, the strong interaction confines quarks to a range of about 3 x 10^{-15} feet (10^{-14} m).

Also, unlike the chargeless photons, gluons carry a special type of charge, called color (no relation to visual colors), which is one of the fundamental properties of quarks. Because they carry color, gluons can change the color of quarks they interact with, and they can interact with each other. According to the theory describing quarks and gluons—quantum chromodynamics (QCD)—there are eight massless gluons.

A CLOSER LOOK

a proton and turned into oxygen. A single proton from the α-particle that had been added to the nitrogen nucleus turned it into the next heavier element.

Meanwhile, Rutherford and many other physicists had been puzzling over another problem. The hydrogen nucleus contains just one proton, and α-particles were found to be the helium nuclei with two positive charges. If the helium nucleus contained two protons, why was its atomic weight four times that of hydrogen? There had to be another heavy particle in the nucleus, but because it was not charged, it could not be detected by the usual methods. Eventually, in 1932, British physicist James Chadwick (1891–1974) devised an experiment to detect and measure the mass of this uncharged particle. It proved to have a mass almost identical to the proton and was named the neutron (see NEUTRONS).

This graph depicts the binding energy of nucleons for all the stable nuclides, as well as for some radioactive nuclides. The curve peaks at around A = 60, indicating that nuclei with atomic mass close to 60 are the most stable.

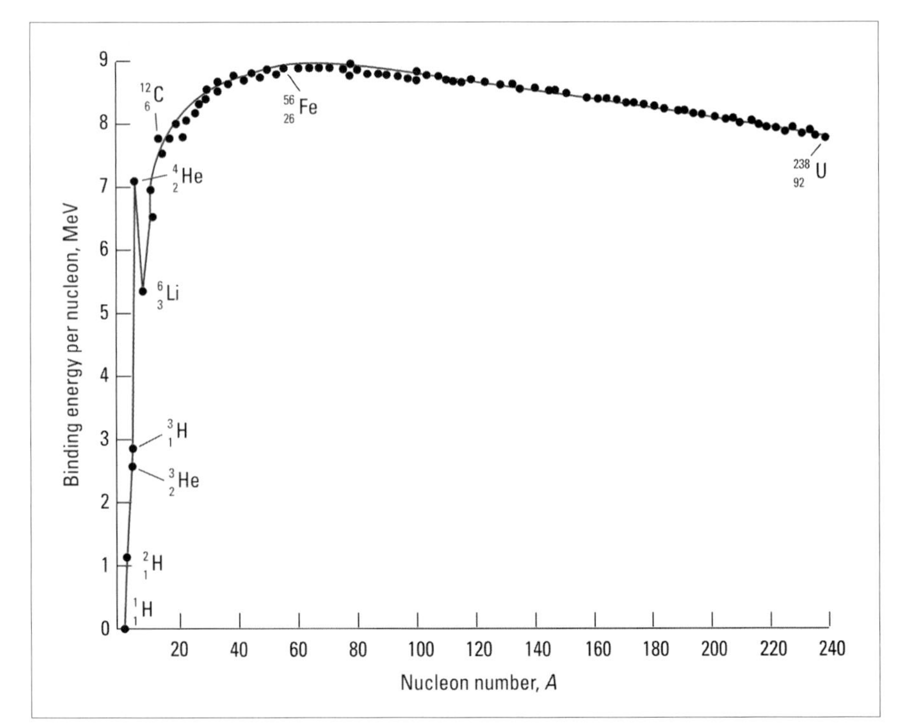

Nucleons and nuclides

Protons and neutrons are collectively known as nucleons, and along with electrons they are the building blocks of atoms (see SUBATOMIC STRUCTURE). The number of nucleons in a nucleus determines the atomic weight of the atom. Protons are 1,836.11 times more massive than electrons, and neutrons are another 0.1 percent more massive than protons. So, even though an atom's nucleus is only 0.0001 times the diameter of the whole atom, it is 4,000 times as massive as all of the electrons surrounding it.

The total number of protons and neutrons in a given nucleus specifies the isotope or nuclide of the element being considered. This number is often called the atomic mass number. It has the symbol A. The number of protons in the nucleus is equal to the number of electrons around it; this is essential information about the chemical properties of an element and therefore defines it. All the atoms in an element have the same number of protons, which scientists call its atomic number (Z). However atoms can have different numbers of neutrons without affecting their chemical properties. These atoms, with differing atomic weights, are called isotopes. For example, the most common isotope of uranium, U^{238}, has a nuclide of 92 protons and 146 neutrons, but U^{235} (used in the first atomic bomb) has 92 protons and only 143 neutrons (see ISOTOPES).

Nuclear forces

Protons and neutrons are tightly packed together in the nucleus. Protons are positively charged—why do they not repel one another and the nucleus not disintegrate? The reason is that all the nucleons are held together by the strong nuclear force. The strong nuclear force is 100 times more powerful than the electromagnetic force, which would cause the nucleus to disintegrate. This strength explains the tremendous energy contained in a nucleus. However, this strong force acts only over a very short distance, no greater than 4 x 10^{-15} in (10^{-11} mm), which is approximately the size of the diameter of a nucleus.

The larger the nuclide, the more important the electric force becomes. To keep the nucleus stable, there must be at least as many neutrons as protons, and with increasing size, even more neutrons become necessary. The largest stable nuclide is Bi^{209}, which has 83 protons and 126 neutrons. With atomic numbers greater than 83, it is impossible to have a stable nucleus, no matter how many neutrons there are, and all larger nuclei undergo radioactive decay (see RADIOACTIVITY).

There is another force, the weak nuclear force that acts between all particles. It is much weaker than the strong force, and it acts over a relatively shorter range. It is, however, strong enough to disrupt some nuclides and cause the nucleus to decay.

Nuclear fusion and fission

Following Rutherford's discovery of the transmutation of nitrogen into oxygen, much research in nuclear physics has sought to alter atomic nuclei in order to create new elements, induce radioactivity, or release large amounts of nuclear energy. Physicists bring about these changes by bombarding nuclei with protons, neutrons, or other particles. Neutrons are easy to work with. As they have no charge, the positive charge of the nucleus does not repel them.

Beginning in 1934, Italian-born physicist Enrico Fermi (1901–1954) found a way of slowing down neutrons by passing them first through water or kerosene. Slower neutrons have a better chance of being absorbed in an atomic nucleus. When a neutron is absorbed by a nucleus, the atomic number remains unchanged, but its mass number increases. So, for example, aluminum (atomic number 13, mass number 27) becomes Al^{28}, the kinetic energy of the neutron being emitted as gamma rays (γ rays; high-energy X rays). The new nucleus that is formed is often unstable and may undergo further spontaneous radioactive decay. For example, in the Al^{28} nucleus, the new neutron changes into a proton, emitting an electron: so aluminum (with 13 protons) transmutes to silicon (with 14 protons).

After achieving a number of atomic changes, Fermi decided to try bombarding uranium (atomic number 92) to see whether he could create an unknown element 93. He believed that he had, but in 1939, German chemist Otto Hahn (1879–1968) and Austrian physicist Lise Meitner (1878–1968) showed that what actually occurred was a splitting of the uranium nucleus with the release of additional neutrons. Among the uranium fragments was the unknown element 43, which was later named technetium.

Fermi had produced nuclear fission (see FISSION). In the course of fission, more neutrons were released, which could each produce yet more fission. It works like a house of cards; one card slips at the top and causes two or three more to slip, and so on until the whole structure collapses very quickly. Physicists soon realized that, under the right conditions, a chain reaction would take place. The chain reaction escalates quickly, releasing enormous energy in the form of an explosion. During the Manhattan Project—the intensive program to produce an atomic bomb during World War II (1939–1945)—scientists succeeded in making bombs from U^{235} and from an isotope of a new element, plutonium Pu^{239}. The plutonium was made by bombarding U^{238} with neutrons, followed by the release of two electrons (see PARTICLE ACCELERATORS). In the course of their experiments, they also produced small quantities of the elements 93 now called neptunium. (The first atomic bomb tested and the second dropped on Japan were plutonium bombs. The first bomb to be used in war was a uranium bomb, a design not tested before.)

As part of the same wartime development, Fermi built the first atomic pile (nuclear reactor) in 1942, at the University of Chicago. Embedding a uranium core in graphite, a type of carbon that absorbs neutrons, Fermi was able to control the chain reaction so that it steadily produced energy without exploding. After the war, physicists improved the design of nuclear reactors and put them to work producing electricity.

Even greater amounts of energy are produced when two hydrogen nuclei crash and are combined to form a helium nucleus—about 24 times more energy than in fission (see FUSION). Accordingly,

In 1938, Enrico Fermi was awarded the Nobel Prize for physics for his transmutation of elements by neutron bombardment.

MAGIC NUMBERS AND NUCLEAR STABILITY

That most stable nuclei have the highest binding energy per nucleon is just common sense: the more energy required to emit a proton or neutron, the less likely such loss will occur, and the longer a nucleus will remain stable. What surprised physicists was that the total binding energy rises dramatically when the nucleus holds certain numbers of protons or neutrons: 2, 8, 20, 28, 50, and 82; also 126 neutrons and (it seemed likely) 114 protons or 184 neutrons. These numbers became known as magic numbers. A nucleus with a magic number of protons or neutrons is very stable; one with a magic number of both protons and neutrons is exceptionally stable.

As understanding of nuclear structure grew, it became apparent that the magic numbers represented the complete filling of nuclear energy levels, much as the inertness of the noble gases reflected completely filled electron energy levels (see NOBLE GASES).

ERNEST RUTHERFORD

Ernest Rutherford was born near Nelson, New Zealand, in 1871. He studied physics there and at Cambridge University in England, where he began to investigate radioactivity. He named two types of radiation, alpha rays (α rays) and beta rays (β rays), and later discovered a third type, gamma rays (γ rays; see RADIOACTIVITY). In 1898 he was appointed professor of physics at McGill University in Canada, where British chemist Frederick Soddy (1877–1956), who later discovered isotopes, worked with him for two years. In 1904, with U.S. scientist Bertram Boltwood (1870–1927), Rutherford developed the earliest technique of radioactive dating.

In 1907 he returned to England as professor of physics at Manchester University, where his assistant was German physicist Johannes Hans Geiger (1882–1945), the inventor of the Geiger radiation counter. It was with an early form of this equipment that Rutherford discovered that alpha rays were particles carrying a double positive charge, which he later identified as helium nuclei.

In 1909 Rutherford suggested to Geiger that he should investigate the scattering of α-particles by gold foil, and it was from the results of these experiments that, in 1911, he was able to announce the existence of the nucleus and to determine its size.

For three years during World War I (1914–1918), Rutherford worked for the British Admiralty on submarine-location methods. Then, in 1917, while his students and colleagues were still engaged in the war, he single-handedly discovered the transformation of nitrogen into oxygen, with the release of a free hydrogen nucleus—which he subsequently called the proton. In 1919 he became professor of physics at Cambridge University and director of the Cavendish Laboratory, where he was in charge of the development of the first particle accelerator (see PARTICLE ACCELERATORS).

Besides his Nobel Prize for chemistry, which he received relatively early in his career, in 1908, Rutherford won many other honors. He was knighted in 1914 and made a lord in 1931—1st Baron Rutherford of Nelson and Cambridge. Rutherford died at Cambridge in 1937.

Johannes Hans Geiger (left) and Ernest Rutherford worked together to investigate the processes that govern radioactivity.

DISCOVERERS

fusion holds even greater promise as a commercial energy source. It is by nuclear fusion that stars produce their immense energy. Fusion is difficult to start and even harder to sustain. An extremely high temperature and an ultradense state of matter—such as was produced in the hydrogen bomb—are needed to slam together the nuclei fast enough to fuse (see BETHE, HANS). The Sun, for example, has a core temperature of about 27 million °F (15 million °C) and a density of about 575 pounds per cubic in (160 g/cm^3)—about eight times the density of gold). Attempts to build a reactor, using a magnetic containment field or lasers to sustain and control hydrogen fusion, have so far had limited success, and fusion power is currently predicted for sometime this century. Fusion power would have other advantages besides efficiency; the fuel for fusion is abundantly available, for example, in seawater, and there would be few decay products to dispose of.

Antimatter and other particles

In 1930 British theoretical physicist Paul Dirac (1902–1984) made a mathematical study of the properties of electrons and protons. He concluded that each subatomic particle should have a corresponding antiparticle, with the same mass but an opposite electrical charge. Within two years U.S. physicist Carl Anderson (1905–1991) discovered the antielectron in cosmic rays, and named it the positron (see ANTIMATTER; COSMIC RADIATION; ELECTRONS AND POSITRONS). Soon after, it was found that certain radioactive isotopes also emitted positrons, when a proton in the nucleus changed into a neutron.

Since then, experiments with cosmic rays or high-energy collisions of particles in particle accelerators have produced other antimatter particles, as well as a growing number of "ordinary" subatomic particles, although most exist for only a tiny fraction of a second (see PARTICLE PHYSICS).

There are two classes of particles. Fermi and Dirac worked out the mathematical rules for one class, and the rules for the other were developed by German-born physicist Albert Einstein (1879–1955; see EINSTEIN, ALBERT) and Indian physicist Satyendra Nath Bose (1894–1974; see BOSONS). The first class are called fermions: they include electrons, protons, and neutrons, together with the neutrino (a remarkable particle with a very tiny, mass) and a number of short-lived particles (see NEUTRINOS).

Fermions are divided into leptons (see LEPTONS), which are affected by electrostatic forces and by the weak nuclear force, and hadrons (see HADRONS), which are affected by the strong nuclear force as well as the electrostatic and weak nuclear forces (see FORCES). To account for the exchange of force and energy between particles, physicists proposed the existence of other exchange particles. These particles are called bosons. The best-known of these is the photon, which was postulated at the beginning of the 20th century to explain the particle-like behavior of light; the others are known as mesons (see LIGHT; MESONS).

NUCLEAR SPIN AND THE MAGNETIC MOMENT OF NUCLEI

Protons and neutrons do not occupy fixed positions in the nucleus. Like the electrons moving around the nucleus, the protons and neutrons occupy energy levels and obey the Pauli exclusion principle, which states that no two fermions can have the same quantum numbers. In addition, each proton and neutron carries its own spin angular momentum. The overall spin of the nucleus is determined by the way the spin and motional angular momenta add together. In many cases, there is complete cancellation, so the nuclear spin is zero, as occurs in the common isotopes helium-4, carbon-12, and oxygen-16. Hydrogen-1, carbon-13, and phosphorus-31 all have nuclear spin of one-half. Higher spins, both integer and half-integer, also occur: Nitrogen-14 has spin 1, and cobalt-59 has spin $\frac{7}{2}$, for example. In a magnetic field, nuclei will fall into an energy level of $(2I + 1)$, where I is the nuclear spin. Thus, spin $\frac{1}{2}$ nuclei will have two energy levels, while nitrogen-14 will have three. The energy difference between these levels falls in the microwave region of the spectrum, and if the nucleus is exposed to microwave radiation, it will absorb only those microwave photons with just the right energy to jump from one energy level to another. This absorption is called nuclear magnetic resonance and is a sensitive indicator of the precise chemical environment of the nucleus.

Using this technique, scientists have been able to learn a lot about the internal structure of molecules and the chemical bonds holding them together. An important medical application is magnetic resonance imaging (MRI), in which specialized magnetic fields and microwave signals are combined to generate magnetic resonance data, which a computer can convert to cross-sectional or three-dimensional images of a patient's internal organs. The technique is so sensitive that it can detect differences in water structure between cancerous and noncancerous cells. It is also valuable because it provides images of the soft tissues of the body, which are almost invisible on conventional X-ray photographs (see X RAYS).

The patient below lies with her head in the circular detector of a magnetic resonance imaging (MRI) scanner during a brain scan.

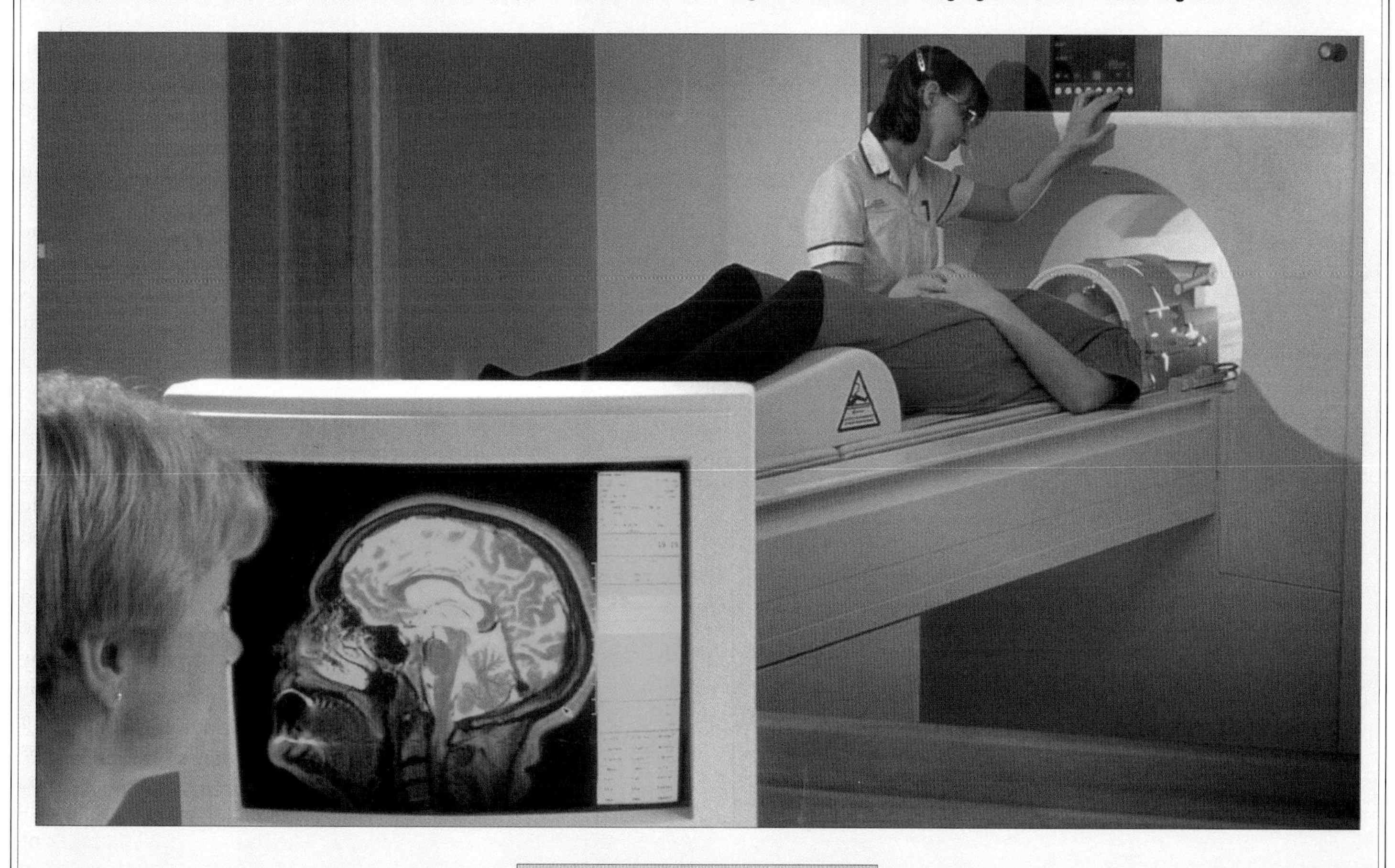

A CLOSER LOOK

Hadrons are now divided into mesons and baryons. In 1964 U.S. theoretical physicist Murray Gell-Mann (b. 1929) proposed that baryons are made up of three even smaller particles called quarks, while mesons consist of two (see QUARKS). Quarks, thought to be the smallest unit of matter, do not seem to exist on their own. They have never been detected in isolation, only bound together with one or two others, illustrating the fact that huge forces would be needed to separate them.

R. SMITH

See also: ANTIMATTER; BOSONS; FISSION; FORCES; FUSION; HADRONS; ISOTOPES; LEPTONS; MESONS; NEUTRINOS; PHYSICS; QUARKS; RADIOACTIVITY; STRING THEORY; SUBATOMIC STRUCTURE; X RAYS.

Further reading:

Basdevant, J. L. 2004. *Fundamentals in Nuclear Physics.* New York: Springer.
Budker, D., and D. Kimball, and D. DeMille. 2004. *Atomic Physics: An Exploration Through Problems and Solutions.* New York: Oxford University Press.

NUCLEIC ACIDS

Nucleic acids are long polynucleotide chains and are the genetic component of every living cell

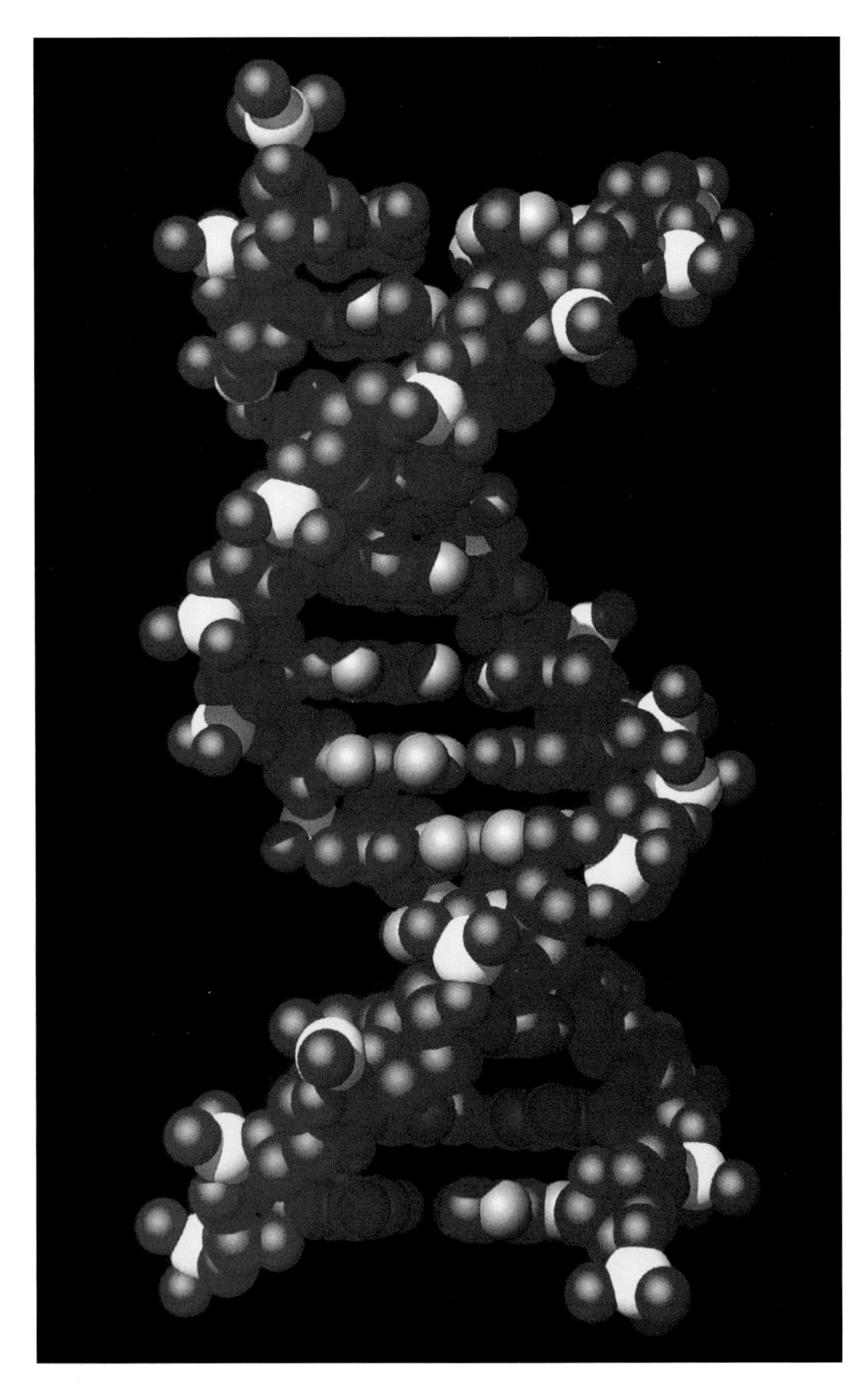

The DNA molecule is composed of two hydrogen-bonded strands of nucleotides. The strands are twisted into a helical shape.

The genetic code that defines who people are and how they grow is incorporated in the nucleic acids: deoxyribonucleic acid (DNA) and ribonucleic acid (RNA). With the exception of blue-green algae and other bacteria, DNA is predominantly found in the nuclei of cells, and RNA is predominant in the cytoplasm. Nucleic acids store information and provide the recipe for the body to synthesize the proteins that are vital to all living organisms. Both DNA and RNA are complex polymers (see POLYMERS). These polymers contain just five different atoms: carbon, oxygen, hydrogen, phosphorus, and nitrogen.

Nucleic acid building blocks

Polymers are large molecules made up of smaller units called monomers. In the case of nucleic acids, the monomer is called a nucleotide. Each nucleotide consists of a nitrogen-containing base, a five-carbon sugar, and a phosphate group. The nucleotides are covalently bonded to one another to form long chains called polynucleotide chains.

There are three nitrogen-containing bases that are common to both nucleic acids: adenine (A), cytosine (C), and guanine (G). DNA has thymine (T) as the fourth and RNA has uracil (U). Cytosine, thymine, and uracil are all derivatives of a class of compounds called pyrimidines. Adenine and guanine are derivatives of a class of compounds called purines (see PURINES AND PYRIMIDINES).

The sugar component of both nucleic acids contains five carbon atoms. In DNA, the sugar is 2-deoxyribose, and in RNA it is ribose. These sugars become bonded to the nitrogen-containing bases through a condensation reaction, creating a nucleoside: for example, adenine gives the nucleoside adenosine (for RNA) and deoxyadenosine (for DNA). The addition of a phosphate group to the fifth carbon of the sugar by means of another condensation reaction yields the nucleotide. The nucleotides then polymerize by forming linkages between this phosphate group and the third carbon of the sugar of the next nucleotide in the chain.

In its native state, DNA consists of two chains of polynucleotides twisted in a double right-handed helix (one that climbs upward in a counterclockwise direction), with the sugar-phosphate chain on the outside and the bases hydrogen bonding together in the center (see picture to the left; see CHEMICAL BONDS). The same pairs of bases always form hydrogen bonds with each other: adenine with thymine and guanine with cytosine. Each complete twist of the helix consists of 10 base pairs. DNA polymers in

CONNECTIONS

- The amino acid sequence in **PROTEINS** is specified by the sequence of **PURINE AND PYRIMIDINE** bases in DNA.

CORE FACTS

- Nucleic acids—deoxyribonucleic acid (DNA) and ribonucleic acid (RNA)—contain the genetic code that is present in all living organisms.
- The monomers of nucleic acids—nucleotides—consist of a nitrogen-containing base, a five-carbon sugar, and a phosphate group.
- DNA consists of two chains of polynucleotides twisted in a right-handed double helix.
- RNA molecules are made up of single polynucleotide chains that may form threads or fold into complex shapes.

human chromosomes have molecular masses in the region of 10^{11}, or 100 billion. RNA molecules, in contrast, are made up of single polynucleotide chains and may fold into complex shapes or remain stretched out as long threads.

The duplication of DNA

To be able to transmit its genetic information, DNA has to be able to duplicate itself. DNA replication begins when certain enzymes begin to unwind the DNA double helix and the two strands come apart. As the two strands are separated, the bases on each strand are exposed and thus free to bind to complementary nucleotides. Because adenine binds only to thymine and guanine binds only to cytosine, the new strand is accurately built. The old strand "directs" the synthesis of a new strand one base pair at a time. This exact replication automatically creates a template for the synthesis of new polymers when the two strands unwind.

The role of RNA

RNA's main function is in controlling the synthesis of proteins (see PROTEINS). RNA is made from DNA by the process called transcription, which is similar to the way DNA duplicates itself with guanine in the DNA pairing with a cytosine in the RNA and vice versa. Thymine pairs with adenine in DNA but in transcription adenine pairs with uracil. The molecule formed is called messenger RNA, or mRNA.

This mRNA guides the synthesis of proteins by units called ribosomes in a process called translation. Ribosomes consist of numerous different proteins and several ribosomal RNA molecules, or rRNA. These rRNA molecules attach themselves to the mRNA, and the bases present in the mRNA determine how the protein is created: starting at one end of the polymer, the mRNA must have the starting code of AUG for protein synthesis to be initiated. A codon is a set of three bases that codes for a single amino acid. The starting code is called the initiation codon. Each successive set of three bases is a specific codon for one of the amino acids, so the mRNA provides the "recipe" for the whole protein synthesis, adding amino acids one by one to the elongating polypeptide chain. When a termination codon is found, it indicates that the protein synthesis should stop.

A third form of RNA, transfer RNA (tRNA), binds to specific amino acids in the cytoplasm and is responsible for transporting them to the mRNA molecule, where they are incorporated into the protein molecule under construction.

S. HOULTON

See also: CHEMICAL BONDS; FUNCTIONAL GROUPS; ORGANIC CHEMISTRY; POLYMERS; X RAYS.

Further reading:

Neidle, Stephen. 2002. *Nucleic Acid Structure and Recognition*. New York: Oxford University Press.

ROSALIND FRANKLIN

Although the discovery of DNA's double helix structure is inextricably linked with the names of British molecular biologist Francis Crick (1916–2004) and U.S. biochemist James Watson (b. 1928), the pioneering work of British biophysicist Rosalind Franklin, shown to the right, was essential in the search for DNA's structure.

Born in London in 1920, Franklin began work at King's College, London, in 1951. Using her skills in X-ray diffraction, she developed a way of hydrating (adding water to) DNA molecules and then controlling the hydration under X rays (see X RAYS). By the autumn of 1952, Franklin had produced diffraction images of sufficiently high quality to put her very close to discovering the structure of DNA. In early 1953 she provided diffraction data that agreed with the structure published by Crick and Watson. However, without her knowledge, one of her colleagues at King's had, a few weeks earlier, showed Watson a copy of her best diffraction photograph, which was sufficiently detailed to enable all the main parameters of DNA to be deduced. One can only speculate how much this photograph helped Crick and Watson in their own studies. Franklin died of cancer in 1958—two years before Crick and Watson were awarded a Nobel Prize for deducing the structure of DNA.

DISCOVERERS

NUCLEOSYNTHESIS

Nucleosynthesis is the production of elements by nuclear reactions in the very young universe and within stars

Most of the elements that make up Earth were made inside the core of a star billions of years ago. The Sun is currently making helium and will produce other elements before it burns out.

The universe started in an explosion termed the big bang, according to scientists, and it has been expanding ever since (see BIG BANG THEORY; UNIVERSE). At the start, about 13 billion years ago, the universe was made of particles of radiation and bore little resemblance to what is now seen. Wherever one looks now, the same chemical elements are present (see ELEMENTS). One of the problems facing mid-20th-century astronomers was explaining the origin of these elements (see ASTROPHYSICS).

Big bang nucleosynthesis

The idea that hydrogen was producing helium inside stars was put forward in 1926 by English astronomer Arthur Eddington (1882–1944). Eddington was working to explain the source of energy in stars. The synthesis of hydrogen to helium would, he rightly theorized, release enormous amounts of energy. During the next decade, European and U.S. astronomers worked to explain the details of this process. Later, in the 1940s, some investigated the idea that elements could have been made at the start of the universe and, in turn, uncovered the origin of the very first elements.

In 1948 Russian-born physicist George Gamow (1904–1968) and two of his students showed that the first elements formed very quickly after the big bang. The elements were produced from protons, neutrons, and electrons—the particles of matter created within the first seconds of the universe (see ELECTRONS AND POSITRONS; NEUTRONS; PROTONS).

Between one second and three minutes after the big bang, when the temperature of the universe had dropped to below one billion K, and the protons (hydrogen nuclei) were fusing with the neutrons to form deuterium nuclei, which then fused to form helium (see FUSION). The first simple elements, termed light elements, had been created.

This fusing of nuclei to produce progressively larger and more stable nuclei is termed nucleosynthesis. Within about a thousand seconds of the big bang, the first era of nucleosynthesis was over. The temperature of the universe had dropped, and it was no longer hot enough for nuclear fusion and, thus, the process of nucleosynthesis to continue. The universe was made of hydrogen and helium with just trace amounts of other light elements, such as deuterium, lithium, and beryllium.

Stellar nucleosynthesis

Hydrogen and helium have always been the most abundant elements in the universe. Within three minutes of the big bang, hydrogen accounted for 77 percent of the universe, and helium 23 percent. These percentages have hardly changed in the intervening 13 billion years. Over 900,000 of every million atoms of the universe are now hydrogen and over 90,000 are helium. Yet, many more elements than the first handful now exist. The third most abundant element is oxygen (673 atoms for every million), followed by carbon (406 atoms). The creation of these elements started later in the universe's history. Such middle-weight elements and heavier ones have been made inside stars within the last ten billion years or so, and their production continues now in the same way.

As the young universe cooled, its material very slowly and very gradually started to come together. Enormous clouds of hydrogen and helium contracted under the pull of gravity to create spheres of material, the stars (see STARS). As a star contracts, the temperature in its core increases. When it reaches about 10 billion degrees, it is hot enough for nuclear fusion, and hydrogen is fused to form helium.

CONNECTIONS

- Almost all the **ELEMENTS** now present on Earth had their origins in the **STARS**.
- Scientists use **SPECTROSCOPY** to find out which elements are inside stars.
- Elements are produced inside stars by nuclear **FUSION**.

CORE FACTS

- The lightest elements, hydrogen and helium, were produced within the first few minutes of the universe.
- Typical stars, such as the Sun, convert hydrogen to helium in their cores and then go on to produce carbon and oxygen.
- The most massive stars with the highest core temperatures produce progressively heavier and heavier elements, up to iron.
- Elements heavier than iron are synthesized inside massive explosions called supernovas.

A star converts its core hydrogen to helium by a chain of reactions known as the proton-proton chain. The process used in more-massive and hotter stars is the carbon cycle. Once the majority of the core hydrogen has been converted to helium, the core contracts, its temperature rises, and carbon and oxygen are produced by a new nuclear process, the triple alpha process. A star like the Sun would typically produce carbon and oxygen but could not fuse heavier elements because its core would not achieve the necessary density and temperature. These would be achieved only by more-massive stars, which would go on to produce the elements neon and magnesium, then, at higher temperatures still, silicon and iron.

A star, which has produced a series of higher and higher elements, has a core made up of layers, like the layers of an onion. Each layer is of a different element, the heavier the element, the closer to the center, with the heaviest of all in the center of the core.

The heaviest elements

Energy is a by-product of the nuclear fusion process. It is produced as all the elements up to iron are formed. Elements heavier than iron cannot be produced by nuclear fusion. Further fusion requires that energy is only achievable in some super massive stars. Stars with at least eight times the mass of the Sun convert the material in their cores by nuclear fusion to higher and higher elements. Iron is produced until the star's core is made of 1.4 solar masses worth of iron. At this point, known as the Chandrasekhar limit, the production of iron and of its by-product, energy, stops (see CHANDRASEKHAR, SUBRAHMANYAN).

The star becomes unstable because there is no outward (energy) pressure to counterbalance gravity's inward pull. Gravity successfully pulls the star in and creates a small, dense neutron star. A shock wave travels out from the neutron and through material left from the original star and triggers nuclear reactions, involving neutron capture, in the stellar leftovers. Within seconds new elements heavier than iron, such as uranium, thorium and cobalt, are created. The star now explodes; it is a supernova (see NOVAS AND SUPERNOVAS). The mechanisms for the production of heavy elements were first described in 1957. Three decades later astronomers got their first chance to make a detailed study of a supernova explosion. Observations of supernova 1987A provided the first confirmation of nucleosynthesis.

Chemical scattering

The elements up to silicon are scattered through space by stellar winds, by old stars pushing off their outer shells, and by nova explosions. The heavier elements are blasted into space by a supernova. The supernova material sweeps up interstellar material as it expands into space. Together they form a hot, glowing cloud called a supernova remnant, which expands indefinitely, scattering its elements through space. These stellar scatterings are the raw material for subsequent generations of stars and planets.

C. STOTT

FRED HOYLE

Englishman Fred Hoyle (1915–2001) was one of the 20th century's leading scientific figures. He tackled and provided solutions to many fundamental astronomical questions. During the 1940s he worked on the production of medium and heavy elements in stars and supernovas, collaborating with, among others, U.S. astronomer William Fowler. A 1957 research paper coauthored with Geoffrey and Margaret Burbidge explained how elements could be produced by a number of different processes, including the s and the r process, which describe how elements heavier than iron are produced. These processes are so named for the rate, slow or rapid, by which neutrons are added to nuclei. The paper, now known simply as B"FH after the four authors, is the basis for present theories of element production.

Hoyle (pictured below), along with Hermann Bondi and Thomas Gold, proposed the steady state theory of the universe in 1948. It described a universe that appears the same at every point and for all time; it has neither beginning nor end. The theory was a rival to the big bang model for about fifteen years.

DISCOVERERS

See also: ASTROPHYSICS; BIG BANG THEORY; CHANDRASEKHAR, SUBRAHMANYAN; ELECTRONS AND POSITRONS; NEBULAS; NEUTRONS; NOVAS AND SUPERNOVAS; NUCLEAR PHYSICS; PROTONS.

Further Reading:

Cowley, C. R. 1995. *An Introduction to Cosmochemistry*. New York: Cambridge University Press.

Pagel, E. J. 1997. *Nucleosynthesis and Chemical Evolution of Galaxies*. New York: Cambridge University Press.

Wickramasinghe, C., G. Burbidge, and J. Narlikar, eds. 2003. *Fred Hoyle's Universe*. Boston: Kluwer Academic Publishers.

OCEANIA

Oceania consists of Australia and more than 10,000 islands scattered across the Pacific Ocean

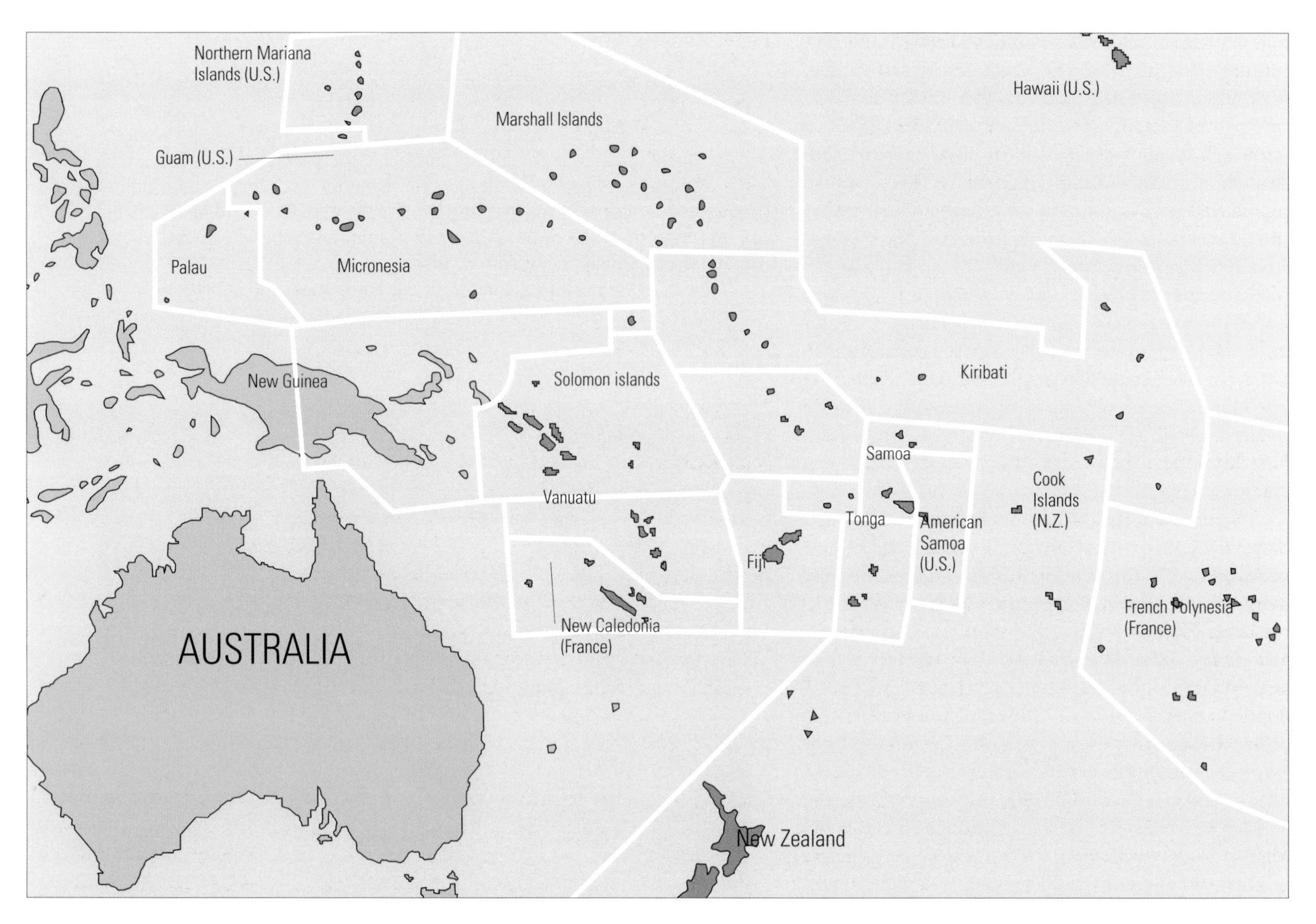

The islands of Oceania stretch across the south and central Pacific Ocean, from Hawaii to New Zealand. The area of ocean controlled by each group of islands is shown by the white lines.

Oceania refers to a the islands dotted throughout the Pacific Ocean (see ISLANDS; PACIFIC OCEAN). Some experts insist that Oceania encompasses even the cold Aleutian Islands and the islands of Japan (see NORTH AMERICA; EURASIA). Disagreement also exists over whether or not Indonesia, the Philippines, and Taiwan should be included in Oceania. Sometimes Oceania includes Australasia (Australia and New Guinea; see AUSTRALIA). However, for the purposes of this article, Oceania is split into three regions: Melanesia, Micronesia, and Polynesia (which includes New Zealand).

Human habitation

Scientists have determined that the more remote parts of Oceania were the last regions on Earth to be inhabited by humans. Anthropologists believe that humans arrived in western Oceania about 40,000 years ago, but they did not reach New Zealand, for example, until 1,000 years ago. It is generally believed that human populations spread from Southeast Asia outward to the Pacific islands.

In 1521, during the voyage of Portuguese explorer Ferdinand Magellan (1480–1521), Europeans made their first contact with the people of these islands. During the next 250 years, explorers, merchants, and treasure hunters from Europe made contact with people on nearly all of the islands of Oceania. By the early 1800s small numbers of Europeans, including sailors, missionaries, and adventurers, lived on most of the larger islands. During World War II (1939–1945), bitter fighting raged throughout the islands of Oceania after they were annexed by the Japanese Empire. The fighting focused major international attention on the region for the first time. Following the war, air transportation linked Oceania with the rest of the world.

CONNECTIONS

- Oceania is made up of many small volcanic **ISLANDS** in the **PACIFIC OCEAN**.

CORE FACTS

- Oceania is the region of more than 10,000 islands that lie in the Pacific Ocean.
- Oceania includes three distinct regions: Melanesia, Micronesia, and Polynesia.
- The evolution of air transportation in the early 1900s increased the economic and political status of Oceania.
- Most of the thousands of islands that make up Melanesia, Micronesia, and Polynesia are thought to be volcanic in origin.

South of Asia

All of the regions now known collectively as Oceania were once popularly called Australasia, meaning "south of Asia." The use of the term *Australasia* resulted in much confusion about exactly what region of the world was being described. In the 21st century, Oceania is treated as a separate region, and Australasia is generally used to refer only to the continent of Australia and the island of New Guinea (although some geographers may also include New Zealand in Australia, not Polynesia). These nations represent the world's smallest continent (Australia) and its second largest island (New Guinea).

Melanesia

Located northeast of Australia, Melanesia is a 2,000-mile (3,220 km) chain of islands that includes Vanuatu, the Solomon Islands, New Caledonia, and Fiji. The total land area of Melanesia is 100,000 square miles (167,000 km^2). The islands are primarily volcanic in origin and are typically surrounded by coral reefs (see CORAL REEFS).

The origins of the people of Melanesia have not been determined definitely, but it is generally believed that the first humans in this region arrived from Asia via a land bridge—an area of land between the continents that was later flooded by rising seas. *Melanesia* means "black islands," named for the dark skin of the inhabitants. Melanesia currently has a population of about 1.25 million people—yet many of the region's islands remain uninhabited.

Micronesia

The approximately 2,250 small islands that make up Micronesia are scattered over an area roughly the size of the United States. Their total land area is only about 1,100 square miles (2,875 km^2) or about the same size as Rhode Island. *Micronesia* means "tiny islands." Located northeast of New Guinea, the region stretches westward from the International Dateline for about 3,000 miles (4,800 km) and extends northward from the equator for about 1,000 miles (1,600 km). Micronesia is divided into several island groups including the Marianas, Marshalls, and Carolines. The deepest ocean trenches in the world are off the coast of the Micronesian islands. For example, the Mariana Trench, located east of the Mariana Islands, plunges to depths of 36,198 feet (11,033 m) below sea level, making it the deepest ocean depth on Earth (see MARINE EXPLORATION).

Micronesia's 260,000 inhabitants occupy only 110 of the islands, with the largest human populations concentrated on the islands of Guam, Saipan, and Tinian in the Marianas. It is believed that the first inhabitants of Micronesia came from the Philippines or Malaysia. Although the population exhibits a wide range of physical features, their culture and appearance are distinct from the inhabitants of Polynesia and Melanesia. Christianity is the dominant religion in Micronesia, and, although 10 distinct languages are spoken, English is understood throughout the region. The islands of Micronesia are predominantly made up of the remains of an ancient volcanic ridge that begins in the vicinity of Japan and runs southward into the region. Nearly all of the islands in Micronesia are surrounded by coral reefs.

Polynesia

The largest region of Oceania is Polynesia (meaning "many islands"), which covers 15 million square miles (39 million km^2) of water. The land area, however, totals only 113,500 square miles (294,000 km^2). Polynesia stretches from the Hawaiian Islands southward to New Zealand and then eastward to Easter Island. Most of the islands are volcanic in origin, and throughout most of Polynesia, the islands are surrounded by coral reefs. Volcanic activity continues in New Zealand and the Hawaiian Islands.

Within these boundaries reside more than four million people, 90 percent of whom live either in New Zealand or on the Hawaiian Islands. There are several theories regarding the origin of the Polynesian people. While a leading theory maintains that these people originated in Asia, other anthropologists point to the possibility of an origin in Melanesia. The similarity of blood types between Polynesians and Native Americans, along with other cultural similarities, seems to indicate that even if the Polynesians did not originate in the Americas, they apparently had extensive contact with the people of these regions at some point in the distant past.

J. HUMPHREYS

See also: AUSTRALIA; CONTINENTS; CORAL REEFS; ISLANDS; MARINE EXPLORATION; PACIFIC OCEAN.

Further reading:

Strathern, Andrew, 2002. *Oceania: An Introduction to the Cultures and Identities of Pacific Islanders.* Durham, N.C.: Carolina Academic Press.

Europeans made their first contact with the peoples of Oceania during the voyage of Portuguese explorer Ferdinand Magellan in 1521.

OCEANS AND OCEANOGRAPHY

Oceans cover over 70 percent of Earth's surface; oceanography is the scientific study of the oceans

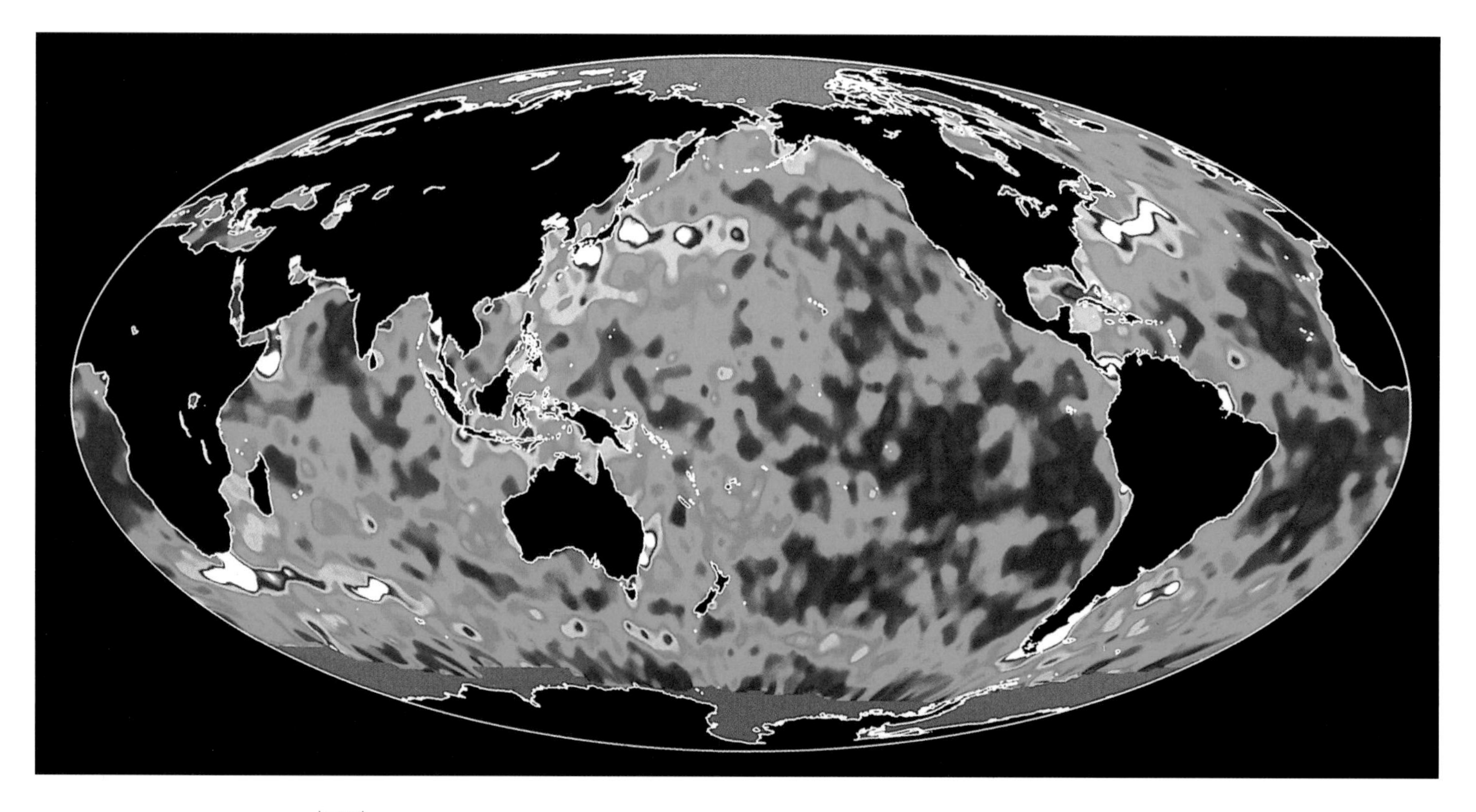

This satellite image highlights the major ocean currents of Earth's principal oceans in yellow, red, and white. Dark blue areas are regions of stable sea heights, such as occur at the centers of oceans.

The oceans are immense, covering over two-thirds of Earth's surface and holding 97 percent of the planet's water supplies. They are also vital to all living things. The oceans serve to regulate Earth's climate, by helping to distribute warmth from the Sun around the globe. They also constitute the main source of the moisture that makes Earth's landmasses habitable. In addition, they provide a habitat for a remarkable variety of living organisms—an estimated 80 percent of all life on Earth.

People have always been fascinated by the oceans. However, in many ways, the oceans remain one of the last wildernesses on Earth that have yet to be explored. Oceanography is the study of the physics, chemistry, geology, and biology of the oceans and the seafloor. In the last century or so, sophisticated techniques have helped oceanographers gain many important insights into the oceans and their effect on the whole planet. Despite their efforts, these vast expanses of water still hold many mysteries.

Origins of the oceans

Some four billion years ago, the slowly forming Earth was a hot, barren planet. However, its ancient atmosphere contained the water vapor that now fills the oceans, which had escaped from deep inside Earth. As the planet cooled, this vapor condensed and fell as rain, draining across the land to collect in hollow basins that then filled to form the oceans. Scientists believe these rains formed the first permanent oceans 3.9 billion years ago. Today this same water continues to circulate between the oceans, dry land, and Earth's mantle and atmosphere, rising from the sea surface and the land into the atmosphere through the process of evaporation and returning to Earth's surface as precipitation (see HYDROGEN; PRECIPITATION; WATER). The size and shape of the ocean basins have been constantly changing since Earth's beginnings owing to the movement of the tectonic plates that form the planet's outer crust (see PLATE TECTONICS).

During the late Paleozoic era and up until about 225 million years ago, scientists believe Earth's dry land was composed of one vast supercontinent called Pangaea (see PANGAEA), surrounded by a single ocean known as Panthalassa. About 200 million years ago, Pangaea broke up, and over millions of years, the pieces continued to drift and formed new continents and change the shape of the oceans. By about 80 million years ago, the four oceans seen today had largely assumed their current size and shape. Today the Pacific is Earth's largest ocean, occupying 64,186,000 square miles (166,884,000 km^2). Next

CONNECTIONS

- In the late 1960s, ocean geologists developed the theory of **PLATE TECTONICS**.
- Scuba diving equipment, developed in the 1940s by French oceanographer Jacques-Yves Cousteau, revolutionized **MARINE EXPLORATION**.

CORE FACTS

- Oceans occupy over two-thirds of Earth's surface. The ocean floor has spectacular features such as mountains, valleys, cliffs, and trenches.
- The water of the oceans is continually circulated by currents, tides, and waves.
- Oceanography has many commercial uses, including in, mining and offshore drilling operations.

in order of size, the Atlantic Ocean is 31,862,000 square miles (82,841,000 km^2). The Indian Ocean is 28,350,000 square miles (73,710,000 km^2), and the Arctic Ocean is 5,427,000 square miles (14,110,000 km^2). In addition, some people define the waters surrounding Antarctica as a fifth ocean, the Southern Ocean.

As tectonic movement continues, the seafloor is spreading in midocean zones where plates diverge; as a result molten magma wells up from below to form new crust. Elsewhere, in places called subduction zones, when two plates collide and one is forced to dive (subduct) beneath the other, a deep trench forms. The crustal material melts down in the mantle and later surges back to the surface to erupt as a line of volcanic mountains, either undersea or on land. Because of the processes of ocean floor spreading and subduction, no part of the ocean floor is more than 200 million years old. (Some continental crust is 3.9 billion years old; see CONTINENTS.)

SARGASSO SEA

The Sargasso Sea is an unusual ocean feature occupying an area of 2 million square miles (5.2 million km^2) in the central Atlantic. This body of water has no shoreline. Instead, it is an eddy bounded by the Gulf Stream to the west, the North Atlantic current to the north, the Canary current to the east, and the North Equatorial drift current to the south (see CURRENTS, OCEANIC). Although these currents surge past, there is little water movement within the Sargasso Sea itself. This static state combined with local winds and the fact that evaporation exceeds precipitation here create a vertical layering of the water. A 3,000-foot- (915 m) layer of light, warm water floats on top of cooler, denser water. At its center, the Sargasso Sea is about 3 feet (1 m) higher than sea level on the North American coast.

High salinity (salt levels) and the steep temperature gradient in the water prevent nutrients from rising upward. As a result, the Sargasso Sea is nearly barren of life. However, the sea does contain eight species of seaweed that float in huge clumps on its surface. The genus of seaweed found here, *Sargassum*, gives the sea its name.

A CLOSER LOOK

Ocean floor

Below the surface, the ocean floor has a wide variety of physical features, including towering mountains, sheer cliffs, and yawning chasms—many of them higher, deeper, or more spectacular than their counterparts on land. In coastal waters, the continents are edged by wide, flat ledges called continental shelves, covered by shallow waters usually no more than 3,300 feet (1,000 m) deep (see CONTINENTAL SHELVES). These shelves often have valleys and low hills similar to those found on land. During ice ages these shelves were dry land because water from the oceans was locked up in the ice forming in the glaciers that covered much of the Northern Hemisphere (see GLACIERS AND GLACIOLOGY; ICE AGES).

At the seaward edge of the shelves, the seafloor drops away steeply in a feature called the continental slope (see CONTINENTAL SLOPES). The foot of the slope is called the continental rise, which consists of a thick layer of sediment, usually more than a mile (1.6 km) thick, that has drifted down to accumulate on the bottom. The extent of continental shelves, slopes, and rises vary between oceans. In general, these features are less prominent in the Pacific than in the Atlantic and Indian Oceans.

Beyond the continental slope, the ocean floor looks very different from any landscape on land. Vast areas are covered by flat featureless abyssal plains of deep sediment and organic matter, called ooze. The deep waters hold some spectacular features of the ocean bed. For example, lines of underwater volcanic mountains called midocean ridges run across the seafloor. These ridges are up to 3 miles (5 km) high and form a near-continuous chain encircling the globe. In most cases, the top of the ridges lie 1 to 2 miles (1.6 to 3.2 km) below the surface. However, in some places they break the water to form islands, such as Iceland, which is located on the Mid-Atlantic Ridge. Midocean ridges are formed by volcanic activity as molten rock surges upward from the mantle to create new oceanic crust (see VOLCANOES).

In 1977 scientists in submersibles discovered an amazing feature at midocean ridges: hydrothermal vents gushing black clouds of hot, sulfurous water (see HYDROTHERMAL VENTS). Seawater entering cracks in the ocean crust is heated by deep-seated magma chambers from which mineral sulfides are leached. These metal-rich waters are then spewed forth from seafloor vents. Tall, chimneylike structures created by deposited minerals are known as black or white smokers.

In addition to the ridges, parts of the ocean floor are occupied by features called rises, such as the Bermuda rise in the western Atlantic. Many of these rises are formed by clusters of underwater volcanoes called seamounts. In warm water the seamounts are capped with coral reefs (see CORAL REEFS).

Seamounts generally do not rise above the surface. Those with a nearly flat top are known as guyots. Scientists believe guyots once lay above or at least near the surface, where their tops were leveled by waves. Later they subsided to an average depth of 3,300 to 6,600 feet (1,000 to 2,000 m) as the volcanic rocks that formed them cooled and contracted. Scientists estimate that the world's oceans hold more than 10,000 seamounts and guyots.

Abyssal hills around 3,300 feet (1,000 m) high are formed when there is not enough sediment to cover up all the volcanic features on the seafloor. These hills are usually parallel to ocean ridges and occupy nearly one-half of the floor of the Pacific. They are far less prevalent in other oceans.

Another feature found worldwide are ocean islands (see ISLANDS). These islands are almost always volcanic in origin, consisting of accumulated lava flows from erupted undersea volcanoes that eventually built up to break the water surface.

In addition to features rising from the seafloor, the deep oceans also hold canyons and trenches descending to enormous depths. Trenches occur in

VOYAGE OF HMS *CHALLENGER*

The year 1872 saw the beginning of the most thorough and scientific ocean study of the 19th century, with the launch of the British research ship HMS *Challenger* (pictured below). For three and a half years, this vessel cruised the globe, while scientists took water samples, temperature readings and depth recordings and sampled marine flora and fauna of all kinds. *Challenger*'s scientists identified more than 4,700 new species and found living organisms at depths of up to 27,000 feet (8,000 m) and thus proved that certain species were capable of living in the deep oceans. *Challenger* seamen recorded a phenomenal depth of 26,850 feet (8,184 m) in the Pacific, having accidentally stumbled upon Earth's deepest point, the Mariana Trench. When the voyage ended, *Challenger* scientists produced a 50-volume report of nearly 30,000 pages. This enterprise gave birth to the modern science of oceanography.

HISTORY OF SCIENCE

subduction zones, where one plate is forced below another. Earth's deepest point, the Mariana Trench in the Pacific Ocean, lies on the boundary between an oceanic and a continental plate. There the seafloor drops away to a depth of 36,198 feet (11,033 m). If Mount Everest, the highest mountain on land, were dropped into this trench, the summit would still lie more than a mile below the surface. Elsewhere, smaller canyons occur at midocean ridges, where tectonic plates pull apart.

Ocean currents

The waters of Earth's oceans are never still but continually circulate in currents, both at the surface and in deeper water. Driven both by the winds and density differences, ocean currents have an important influence on climates worldwide (see CURRENTS, OCEANIC). Currents push warm water from the tropics toward the north and south, moderating winters in temperate and polar latitudes, while cold currents originating in polar waters cool regions in lower latitudes. In many ways, ocean currents are like huge rivers, sometimes with well-defined boundaries and regular patterns of flow. However numerous smaller flows called meanders and eddies periodically break away from the main currents, sometimes circling around to rejoin them at a different point.

The Gulf Stream is one of the world's best known surface currents (see GULF STREAM). Leaving the Gulf of Mexico via the Florida Straits, this warm current flows north along the east coast of North America at speeds up to 130 miles (220 km) per day. Swinging away from the coast at Cape Hatteras, North Carolina, the warm water, now the North Atlantic

Current, flows across the Atlantic to warm the shores of western Europe. Without this warm current, winters in western Europe would be far more severe than they are today.

Surface currents are mainly driven by the wind. Bent by Earth's rotation and the shapes of continents, these currents swirl around in giant circles called gyres that flow clockwise in the Northern Hemisphere and counterclockwise in the Southern Hemisphere (see CORIOLIS FORCE).

Surface currents also help drive two other important forces in ocean circulation—convergence and divergence. Where currents converge, surface water collects in such abundance that some of it sinks. Where currents diverge, their movement pulls water up from the depths. This downwelling and upwelling of water can extend to depths of up to 3,000 feet (915 m).

Waves and tides further help the water in the oceans to circulate (see TIDES; WAVES AND WAVE ACTION). Waves are caused by the action of winds that blow over the ocean surface and create ripples that intensify to form waves. Tides are regular changes in water level, mainly caused by the tug of the Sun's and Moon's gravity on the oceans.

Shaping coasts

Shorelines take many forms throughout the world, but all fall into two general categories: erosional and depositional coasts (see COASTS). Erosional shorelines are shaped by the action of waves, which hurl sand and pebbles against the shore and wear it away. This action frequently creates cliffs and a variety of coastal features, depending on the geology of the rocks and their ability to resist the waves (see CLIFFS; EROSION). Where coastal rocks are soft, the waves may undercut a cliff, which moves steadily inland as sections collapse into the sea. Where more-resistant rocks occur, wave action may create holes and caves and eventually arches and isolated pillars called stacks.

Beaches on eroding shorelines are generally composed of materials removed from cliffs or coastal rocks by the waves. Wave action then sorts these materials by size and density, leaving the heavier fragments up the beach while carrying smaller pieces and sand out to the intertidal zone. This area of the

A space shuttle picture of open ocean turbulence in the Caribbean between the Panama Canal and Jamaica. The interaction between two adjacent spiral eddies has produced a number of smaller spin-off eddy currents.

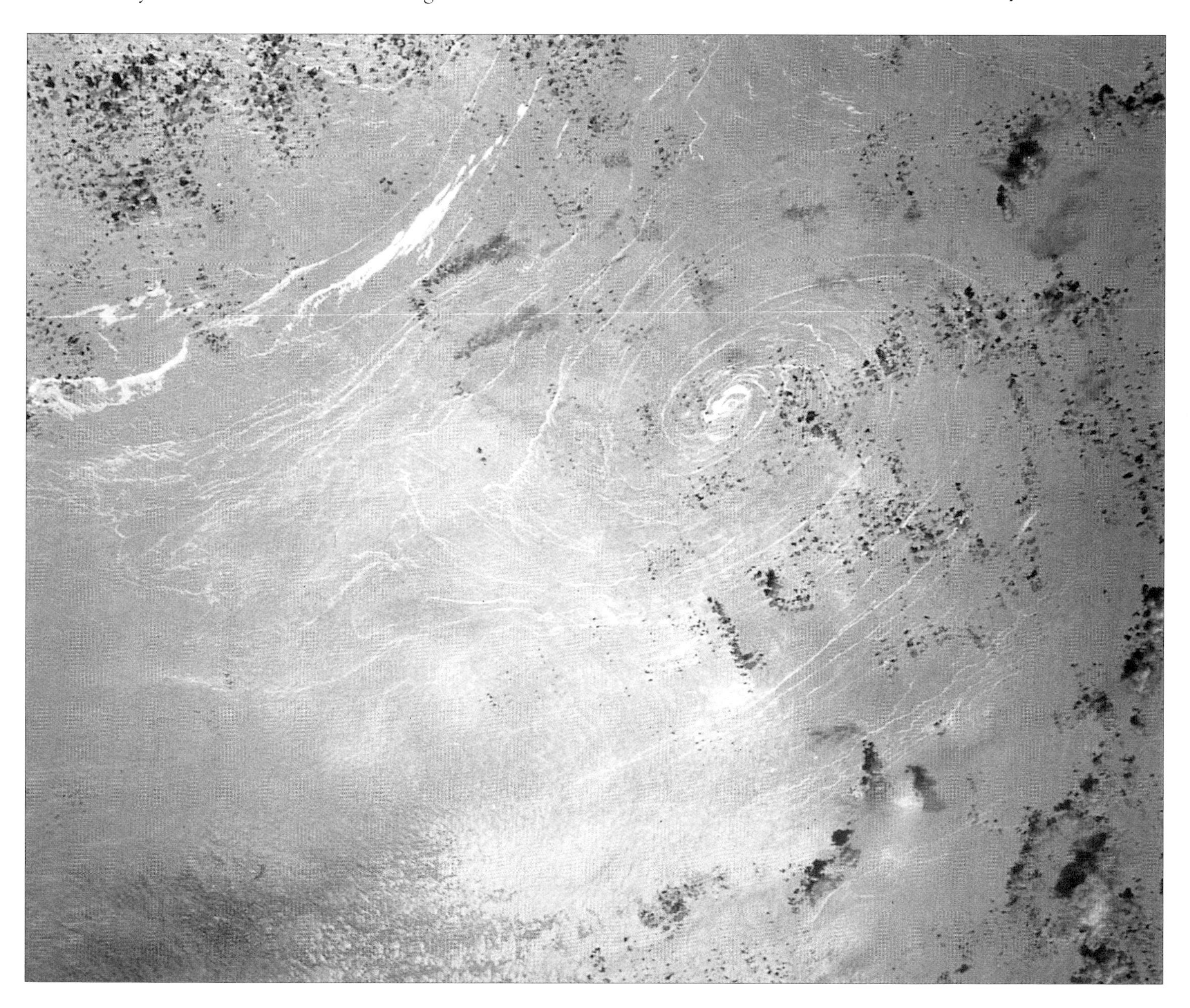

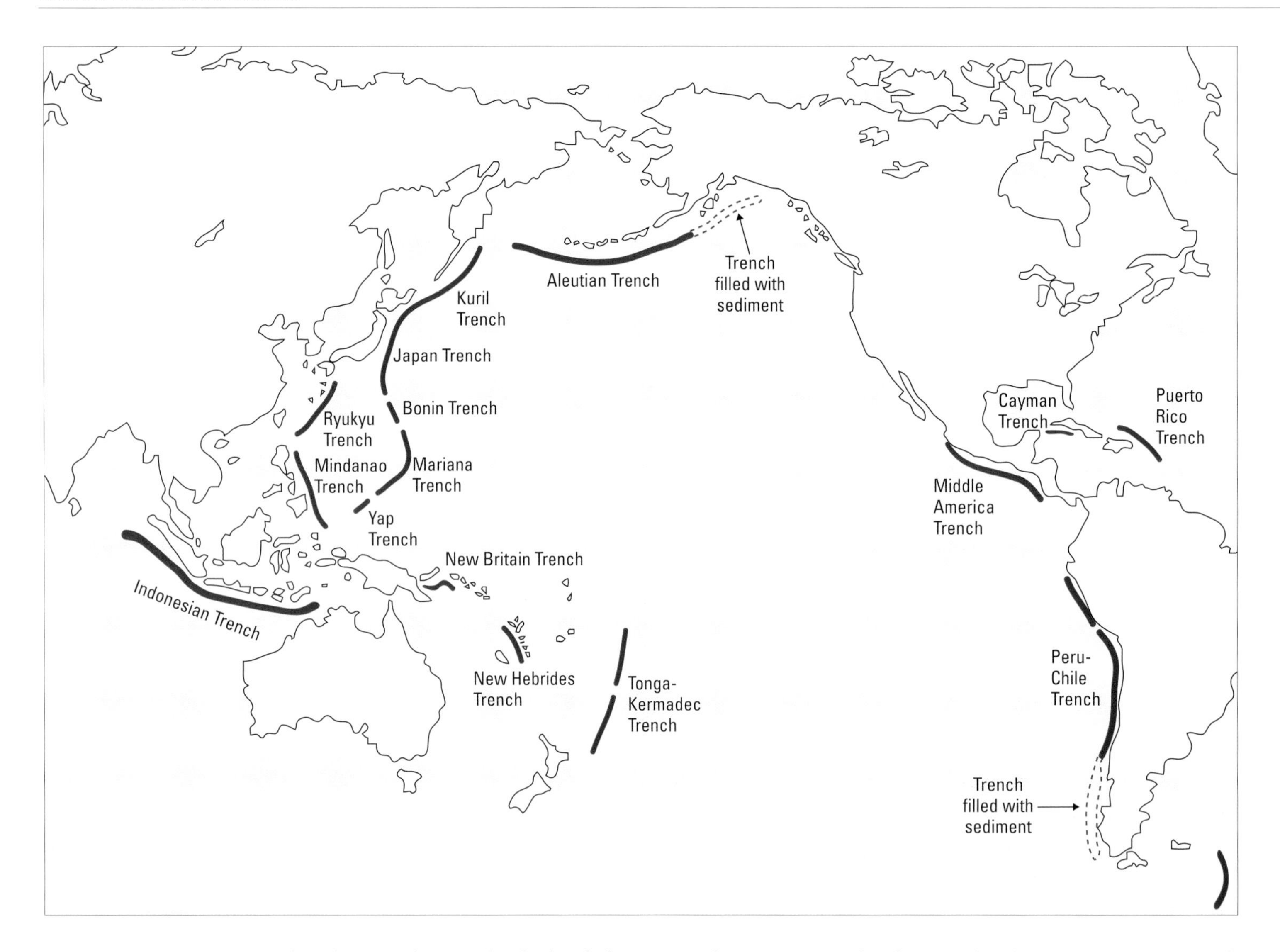

Most seafloor trenches occur in the Pacific Ocean. These trenches mark places where plates of lithosphere have plunged down into the mantle.

shoreline is submerged at high tide but exposed at low tide (see TIDES).

Sometimes near-shore currents carry eroded material along the coast, where it is added to depositional coastlines. These shorelines are mainly composed of sediment, which often originates far inland and is carried to the sea by rivers. Sediment is then deposited at the coast to form features such as deltas, spits, and barrier islands (see DELTAS; BARRIER ISLANDS). A process called longshore drift occurs where waves strike the shore at an angle that allows them to transport sediment along the coast. As a result, a beach may be made of sand carried from a river mouth many miles away. Depositional shorelines may also be formed by materials originating offshore. For example, the white sand beaches of the tropics are often made by the erosion and weathering of coral skeletons offshore (see WEATHERING).

Oceanography

Oceanography is the study of the oceans and ocean floors. This broad science is broken down into four main disciplines: marine geology and biology and chemical and physical oceanography. Chemical oceanographers analyze the composition of seawater, which contains more than 80 naturally occurring elements. The most common mineral dissolved in seawater is sodium chloride—common salt, which accounts for the water's salty taste. Some chemical oceanographers are involved in finding ways to extract valuable metals from seawater.

In recent years, chemical oceanographers have developed computer programs to model the circulation of natural and human-made chemicals through the oceans. Chemical oceanographers also study the effects of carbon dioxide and global warming on the oceans (see GLOBAL WARMING).

Physical oceanographers concentrate on the circulation and physics of water in the oceans, particularly currents, tides, and waves. Their findings have many practical applications, including working with engineers to prevent shoreline erosion.

Marine geologists study the rocks of the ocean floor together with the sediments that settle on them. Much of this work is done aboard ships, using sonar to detect submarine features such as mountains and canyons and powerful drills to extract seafloor samples. In the late 1960s, marine geologists developed the theory of plate tectonics, which revolutionized geology (see GEOLOGY; GEOPHYSICS). Geologists investigating the Atlantic seabed discovered that the youngest rocks lay at the mid-ocean ridge in the center, with progressively older rocks on either side. This discovery suggested that new crust was being created along a plate boundary here, causing the seafloor to spread. Marine geology also has many commercial uses. For example, major energy corporations employ these scientists to search for deposits of oil, gas, and metal deposits on the seafloor (see MINING AND PROSPECTING).

Marine biologists focus on the flora and fauna of the oceans. They distinguish two main ocean habitats: the neritic zone, or coastal waters, and the open oceans. The open ocean is further divided into several vertical layers, each of which has different conditions of light, oxygen level, and temperature. Each of these conditions supports different organisms. The warm, sunlit waters of the euphotic (upper) zone extend down to 660 feet (200 m). The upper part of the bathyal zone, mid-depth between 660 and 6,600 feet (200 and 2,000 m), also contains some light, but none reaches the pitch-black, ice-cold abyssal zone below 6,600 feet (2,000 m).

In the late 1970s the discovery of "black smokers" on the ocean floor created the new specialty of submarine hydrothermal biology. Hydrothermal vents caught the interest of biologists because of the wide range of highly unusual creatures they support, including tube worms, white crabs, and giant clams. Unlike all other living things on Earth these organisms do not derive energy from the Sun. Instead they depend on bacteria that thrive on the sulfurous metal-rich chemicals released by the vents. In recent years many marine biologists have become increasingly concerned about the detrimental effects of human activities on ocean life. Many blame pollution, overfishing, and other factors for whale beachings and dwindling stocks of many marine species worldwide (see POLLUTION).

History of oceanography

The term *oceanography* was coined in 1859. However the Greek philosopher Aristotle (384–322 B.C.E.) is often credited with having founded the science long before, in the 4th century B.C.E. Aristotle classified dozens of marine creatures, including sponges, clams, and oysters. He distinguished between viviparous organisms, which bear live offspring, and oviparous organisms, which lay eggs. He also separated warm-blooded, air-breathing marine mammals from cold-blooded, gill-using fish. Aristotle may have commissioned his student the Macedonian emperor Alexander the Great (356–323 B.C.E.) to collect samples for him. In 333 B.C.E. Alexander ventured into the sea in a primitive diving bell, which trapped an air bubble in its dome for the diver to breathe (see MARINE EXPLORATION).

In 1855 the first scientific study of physical oceanography was published. The work of U.S.naval officer Matthew Fontaine Maury (1806–1873), it was entitled *The Physical Geography of the Sea*. As superintendent of the U.S. Navy Depot of Charts

The primitive diving bell of Alexander the Great, who may have provided ocean specimens for his tutor, Greek philosopher Aristotle.

OCEANOGRAPHERS PREDICT THE WEATHER

Meteorologists who study climate have long known that the oceans have a profound effect on weather (see CLIMATE; WEATHER). Until the late 20th century oceanographers could not see enough of the oceans to understand global circulation patterns. The introduction of satellite monitoring in the 1970s meant that the global picture could be obtained. Data supplied by satellites on surface winds, ocean temperatures, colors, and other factors is sent by radio waves to supercomputers on Earth, where it is processed to make color-coded maps or other geographic information system (see GIS). Researchers then use computer models to predict weather patterns. Programmers feed data on past ocean and weather patterns into the computer, which uses the information to predict how current weather patterns will develop when it receives new data.

Several nations are now cooperating on long-term projects to study the links between ocean circulation and weather. Scientists participating in the World Ocean Circulation Experiment discovered that periodic droughts in the tropics were linked to an interaction between the ocean and the atmosphere called El Niño (see EL NIÑO). A related project, the Tropical Ocean Global Atmosphere program, is studying seasonal variations in the tropics and the effects of El Niño in the eastern Pacific.

SCIENCE AND SOCIETY

and Instruments, Maury received data from navigators traveling all over the world, which he used to describe effect of currents and winds.

Until the late 19th century much of what lay below the ocean surface remained a mystery. Most scientists were convinced that the deep oceans were devoid of life because they believed animals could not survive the high pressure of great depths. Then, in the 1860s, U.S. geologist G. C. Wallich dredged worms and mollusks from a depth of 4,000 feet (1,200 m). From 1872 to 1876, the British research vessel HMS *Challenger* conducted the first scientific expedition devoted to study of the oceans. *Challenger* circled the globe, while scientists on board made numerous important discoveries, including the occurrence of manganese nodules on the seabed (see box on page 1088).

Undersea exploration

The 20th century saw many important advances in technology that helped unlock the mysteries of the oceans. Among them was sonar, which uses sound waves to accurately measure ocean depths (see SOUND). Sonar was developed during World War I (1914–1918) to detect enemy submarines. Modern research ships using sonar direct pulses of sound at the seafloor and monitor the time taken for the echoes to return. The technique was used to compile the first accurate charts of the floor of the deep oceans. Sophisticated sonar systems such as GLORIA—short for geological long range inclined asdic (sonar)—towed by a research vessel, can survey 7,700 square miles (20,000 km^2) of seafloor per day.

The ancient Greeks, including the Macedonian emperor Alexander the Great (356–323 B.C.E), had experimented with diving bells but could spend only a short time under water. Dives grew considerably longer in the early 1800s, with the introduction of diving suits and air supplies. However, these early diving suits were cumbersome and the metal helmets were heavy. In addition, the hose that supplied air from the ship limited the diver's range. In the 1940s French oceanographer Jacques-Yves Cousteau (1910–1997) helped to invent scuba (self-contained underwater breathing apparatus) equipment. This invention set divers free of hoses. Scuba tanks mounted on a diver's back allows the diver to swim freely under water.

In the 1940s divers were still limited by water pressure to depths of about 165 feet (50 m). To explore greater depths, engineers of the 1950s began to develop small, one- and two-person submersibles. In 1960 Swiss naturalist Jacques Piccard (b. 1922) explored the depths of the Mariana Trench in a submersible called a bathyscaphe. Deep submergent vehicles (DSVs) such as the U.S. Navy's *Alvin*, launched in 1964, and the Russia submersible *Mir 1* now regularly descend to depths of 4 miles (6.5 km) or more, carrying out missions of both scientific and economic importance (see GEOPHYSICS). Scientists on board photograph, film, and collect samples of all kinds using cameras and robotic arms.

In 1985 a team led by U.S. oceanographer and undersea explorer Robert Ballard (b. 1942) located the wreck of the ocean liner *Titanic* in the North Atlantic. This famous vessel, once deemed unsinkable, sank after colliding with an iceberg in 1912, and 1,490 people drowned. This discovery was made with an unpiloted submersible called a remote operated vehicle, or ROV. These vehicles also carry out routine tasks such as repair and maintenance work on oil rigs and underwater pipelines and cables.

Modern oceanography

In the second half of the 20th century, oceanographers were aided by increasingly sophisticated technology. Unfortunately, with this technology the cost of ocean studies soared. Research became concentrated among a few large institutions, such as Woods Hole Oceanographic Institution in Massachusetts, and Scripps Institution of Oceanography in California. In recent years ocean research has shifted from exploring the seas just to see what is there to finding answers to specific questions, often of economic or ecological importance

One such project is the Ocean Drilling Project, which involves 20 countries. It was created to collect samples from the ocean floor. Researchers aboard the program's drilling ships study the geological history of the ocean basins and the structure of tectonic plate boundaries. The drilling ship, *Glomar Explorer*, can reach depths of more than 30,000 feet (9,145 m). The Ridge Inter-Disciplinary Global Experiments focus specifically on the dynamics of the spreading seafloor (see PLATE TECTONICS). Data supplied by this project helps to predict undersea volcanoes, earthquakes, and the resultant tidal waves (see EARTHQUAKES; VOLCANOES; WAVES, TIDAL.)

At least two major international studies are assessing the impact of pollutants on ocean life, including the burning of fossil fuels. The Joint Global Ocean Flux Study is attempting to assess the impact of global warming (see GLOBAL WARMING). Participating scientists measure phenomena such as algae and plankton blooms, which increase with rising temperatures. The Global Ocean Ecosystem Dynamics Program is looking at the causes of variations in populations of marine animals.

Today much of the data used by oceanographers is supplied by satellites, which study the oceans from far above. The National Oceanic and Atmospheric Administration (NOAA; see NOAA) began using satellites to monitor the oceans in the late 1960s. Sensors aboard satellites measure factors such as salinity (salt levels; see SALT), sea surface temperatures, distribution of algae and other plankton, and wave and tide flow (see REMOTE SENSING). Infrared images reveal warm and cold currents.

From the late 1980s oceanographers began to utilize observations made by astronauts aboard the space shuttle. The astronauts were able to pick out details not visible close up. For example, a calm sea shines more in sunlight than a turbulent one does. In 1984 photographs taken by astronauts overturned the long-held belief that ocean eddies were rare. These spiral currents or whirlpools that can cover huge expanses of ocean are in fact common and can survive for years. Astronauts now receive training in oceanography and are often assigned ocean studies during their missions.

J. GREEN

See also: BARRIER ISLANDS; CLIMATE; COASTS; CONTINENTAL SHELVES; CONTINENTAL SLOPES; CONTINENTS; CORAL REEFS; CORIOLIS FORCE; CURRENTS, OCEANIC; DELTAS; EARTHQUAKES; EL NIÑO; EROSION; GEOLOGY; GEOPHYSICS; GIS; GLOBAL WARMING; GULF STREAM; HYDROTHERMAL VENTS; ISLANDS; MINING AND PROSPECTING; NOAA; POLLUTION; PRECIPITATION; REMOTE SENSING; SOUND; TIDES; VOLCANOES; WATER; WAVES, TIDAL; WAVES AND WAVE ACTION; WEATHER; WEATHERING.

Further reading:

Garrison, Tom. 2005. *Oceanography: An Invitation to Marine Science*. Belmont, Calif.: Thomson Brooks/Cole.

Thurman, Harold V., and Alan P. Trujillo. 2001. *Essentials of Oceanography*. Upper Saddle River, N.J.: Prentice-Hall.

SYLVIA A. EARLE

Sylvia A. Earle (pictured right in the 1970s) is one of the world's foremost deep-sea explorers. Born in 1935, she first ventured into the deep as a 17-year-old student of the then-new sport of scuba diving. Earle graduated in 1955 and received her doctoral degree from Duke University, North Carolina, in 1966. Her doctoral thesis on the biology of algae was a groundbreaking study of marine life.

Earle made history in 1970 when she led a team of five female scientists on a two-week mission in an undersea habitat. The team inhabited two massive tanks joined by a passageway, submerged in 50 feet (15 m) of water off the U.S. Virgin Islands. The project, called Tektite II, was designed to see how well humans would perform in an isolated submarine environment. In the event, Earle and her colleagues spent most of their waking hours outside the shelter, exploring their undersea neighborhood. They cataloged 153 species of plants, 26 of which had not been recorded in the Virgin Islands.

In 1979 Earle accomplished her most daring feat. She plunged alone to a depth of 1,250 feet (375 m): lower than any human has ever dived. Earle made her record-breaking dive in a pressurized suit called a Jim suit. The dive was extremely risky: one tiny hole in the suit and she would have been crushed. However, all went well, and Earle spent three hours walking on the seabed, collecting specimens with Jim's robotic arms.

Earle longed to share her glimpses of the undersea world with others. In 1982 she and engineer Graham Hawkes founded Deep Ocean Engineering, an organization that designs and builds submersibles. The company created the one-person submersible called *Deep Rover* and also *Deep Flight*, a vessel designed to "fly" through the ocean at depths greater than 4,000 feet (1,200 m).

In the early 1990s, Earle acted as chief scientist of the National Oceanic and Atmospheric Administration (NOAA), the first woman to be appointed to this position. She is now explorer-in-residence at the National Geographic Society. A prolific author and ardent environmentalist, she has written over 125 scientific and popular publications, as well as lecturing and continuing her work on submersibles.

DISCOVERERS

OIL

Oil is a fossil fuel formed from the remains of ancient organisms

Most U.S. offshore oil production rigs are located in the Gulf of Mexico. The extracted oil is transported to refineries beside the Mississippi River close to New Orleans and Baton Rouge, Louisiana.

CONNECTIONS

- Crude oil has virtually no intrinsic value, but the products made from it, including **EXPLOSIVES**, gasoline, soaps, fertilizers, **DETERGENTS**, and **PLASTICS** have made it invaluable.

Life in the 20th century has been shaped more profoundly by oil than by any other single substance. Oil-derived products have fueled the enormous postwar industrial expansion, helped bring heat and electricity to virtually the whole industrialized world, and spawned numerous new industries, such as the manufacture of plastics and detergents (see DETERGENTS; PLASTICS). Demand for oil in the 21st century continues to exert a major influence on geopolitics.

The existence of petroleum and natural gas (see NATURAL GAS) has been known since ancient times. People who drilled for saltwater often came across the gas, and it was sometimes burned to evaporate the brine. Oil obtained from natural seepages had other uses, for example, as bitumen in building, in pitch for ships, and in various medicinal applications. Indeed, it was the use of oil by the Seneca Indians of Pennsylvania (Seneca oil) that inspired the term *snake oil,* a substance sold as a medicine at traveling medicine shows. By the end of the 18th century, Burma had more than 500 wells producing about 40,000 tons (35,700 metric tons) of oil per year. By the end of the 19th century, millions of tons of oil were being extracted annually from Baku in Azerbaijan and Titusville in Pennsylvania.

Despite these early exploits, however, it was not until the invention of electric lighting and the internal combustion engine—the core of the automobile—that oil became the world's predominant fuel.

The early-20th-century surge in demand—and corresponding opportunity for profit—prompted the invigorated efforts to discover new deposits. The wealth generated through oil led to its nickname, black gold.

The uses of oil expanded considerably throughout the 20th century as more and more applications were discovered. Today petroleum products are used as transportation fuels (around half of total use) and as fuels for electricity generation, as well as for manufacturing soaps, paints, plastics, fertilizers, petroleum gel, solvents, and explosives (see ELECTRICITY; PETROCHEMICALS).

Origin of petroleum

The word *petroleum* literally means "rock oil" and refers to its occurrence in the pores of sedimentary

CORE FACTS

- Oil is the liquid form of petroleum, a complex mixture of hydrocarbons formed from the remains of ancient animals and plants. It is commonly found underground in the pores of sedimentary rocks.
- The raw material produced by drilling is called crude oil, and it can range from a pale yellow volatile liquid to a black viscous liquid or a semisolid.
- Oil is recovered by drilling into the reservoir rocks and bringing the raw product—crude oil—to the surface.
- Crude oil undergoes a relatively extensive refining process to produce useful petroleum products.
- Around half of all petroleum products are used as fuel for transportation; other uses include electricity generation and the manufacture of a wide variety of products such as soaps, paints, plastics, fertilizers, petroleum gel, solvents, and explosives.

rocks (see SEDIMENTARY ROCKS). Like coal and natural gas, oil is a fossil fuel formed from the remains of ancient organisms such as algae and plankton. This original organic material decayed and became embedded in fine-grained muds. Over time the succeeding layers buried the lower layers deep underground, where a series of physical and chemical changes took place.

The hydrocarbons in petroleum can be traced back to two basic pathways from the original organic material (see HYDROCARBONS). Around 10 to 20 percent comes from the hydrocarbons already existing in the body of organisms; the remaining 80 to 90 percent is derived from the conversion of lipids, proteins, and carbohydrates into kerogen, the organic material dispersed in sedimentary rocks. For both pathways, the biological, chemical, and physical activity (known as organic diagenesis) alters the original material without any need for profound temperature change (see GEOCHEMISTRY).

The next major phase of petroleum formation is called catagenesis, which is the alteration through the action of heat of the original organic material into petroleum. Under the second pathway described above, the high temperatures crack the kerogens and release hydrocarbons, producing another intermediary product, bitumen.

Petroleum geology

The large subsurface petroleum accumulations found today are far too big to be the original site of the petroleum. Over time, petroleum deposits migrate from source rocks to reservoir rocks. Since extraction is such an expensive operation, petroleum prospectors have to understand what these conditions are to ensure that their efforts focus on the most promising areas (see MINING AND PROSPECTING). The science of petroleum geology has advanced rapidly to investigate these conditions (see GEOPHYSICS; SEISMOLOGY).

The various elements that make up petroleum production can be thought of as a system, comprising source rocks, migration paths, reservoir rocks, seals, traps, and the geological processes that produce each of them. To reduce the risk of drilling expensive "dry holes," geologists have to assess each element of a petroleum system.

Petroleum undergoes three main types of migration: primary migration, which is movement within the fine-grained portion of the mature source rock; secondary migration, which is any movement outside the source rock; and tertiary migration, which is the movement of a previously formed accumulation.

Location of oil reserves

Discussion of oil supplies focuses on proven reserves—those deposits that have been identified as economically viable to extract. The estimated one trillion barrels of reserves are very unevenly distributed throughout the world. The Middle East and Africa contain approximately 700 million barrels, North, Central, and South America some 118 million barrels, eastern Europe and Siberia 68 million barrels, and East Asia and Oceania 57 million barrels. (1 barrel equals 42 gallons; 159 liters).

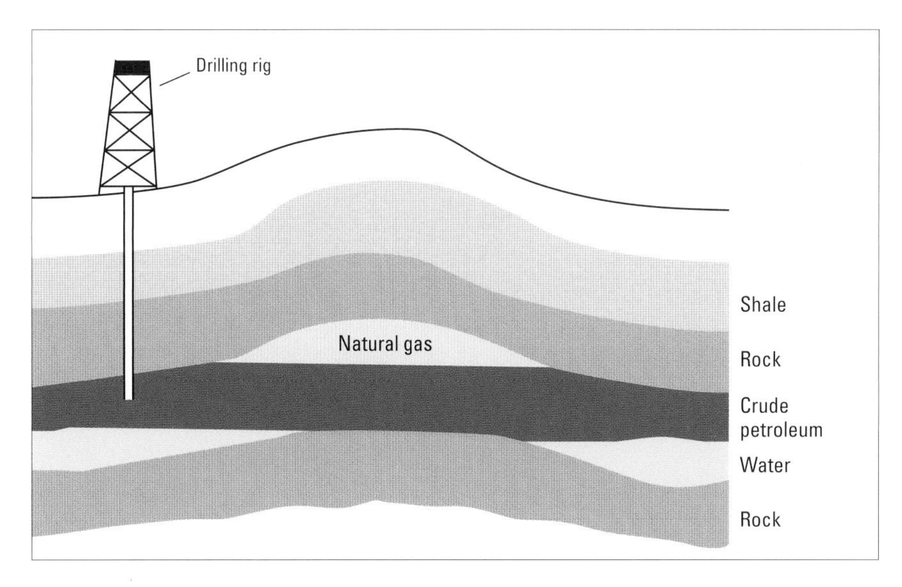

Crude petroleum deposits lying between rock layers can be extracted by drilling through the layers from an overhead oil rig.

Composition of oil

All crude oil is made up almost entirely of hydrogen and carbon. These are combined in hundreds of different ways, however, making crude oil derived from different sources very diverse. A typical crude oil will contain four basic types of hydrocarbon molecules: paraffins (or alkanes), open-chain molecules with single bonds between the carbon atoms; naphthenes (cycloalkanes), characterized by alkane rings; olefins (alkenes), with one or more double bonds between the carbon atoms; and aromatics (arenes), hydrocarbons with one or more benzene rings (see FUNCTIONAL GROUPS). These hydrocarbons can be processed to produce molecules of any type and size. In addition to these compounds, crude oil contains small amounts of nonorganic substances (usually less than 3 percent) such as nitrogen, sulfur, oxygen, and trace metals. The percentage of inorganic content in a given crude oil also varies.

TEXAS, THE OIL STATE

In the United States, people are reminded of oil when they think of Texas. Texaco, one of the major international oil companies, began its existence as the Texas Company. Texas was also one of the world's original places of heavy oil production. The discovery of Spindletop in 1901 is considered the greatest find of all time. The first oil well flowed 80,000 barrels a day, which amounted to half the total U.S. production at that time.

An oil rush began within days of this discovery. Prospectors bought up land at very high prices and began drilling. The euphoria was relatively short-lived, however, because within a year the entire Spindletop reservoir was producing only around 5000 barrels a day.

The Gulf coast area of Texas remains one of the world's chief refining centers. Its importance is reflected in the use of West Texas Intermediate oil as a "marker crude oil." The price of this oil is a regional price against which other crude oils are compared in international trading.

A CLOSER LOOK

A jack pump, or "nodding donkey," is used to extract oil where the natural pressure in the oil well is too low to force it out at an economical rate.

Petroleum processing

Crude oil has virtually no intrinsic value. Petroleum products are what make it an important material. Crude oil must undergo a relatively complex processing operation before it is transformed into useful products. Once the crude oil has been extracted, it is transported to a refinery. Since the distances between the major producing regions and the major refining regions is large, around 60 percent of crude oil is transported by sea. North America has the largest and most sophisticated refining industry (see REFINING).

DEPLETING OIL SUPPLIES

Many people became worried about the scarcity and geographic concentration of oil reserves during the oil crises of the 1970s. In particular, the power of the Organization of Petroleum Exporting Countries (OPEC)—an organization dominated by Middle Eastern countries—to control production and set prices provoked many Western countries to seek greater energy independence.

Since then, countries have implemented energy conservation policies and programs to diversify their energy mix, and other energy sources, particularly natural gas, have become popular. Nevertheless, the industrialized world continues to depend heavily on oil, and with rapid industrialization underway in Asia, people think that the demand for oil will increase significantly.

For these reasons, the adequacy of oil supplies to meet this demand is again a topic of considerable interest among academics, government policy makers, and industrialists alike. The world's requirement for oil currently amounts to about 80 million barrels per day. The discovery and development of new reserves is not expected to match this rate, but it seems supplies should meet demand until about the year 2035.

LOOKING TO THE FUTURE

Fractional distillation

The first stage of the refining process is the physical separation of the components of crude oil. The oil is divided into various products of different weights through heating. This stage is called fractional distillation. The basic categories include light products such as gasoline, naphtha (used to manufacture chemicals), and petroleum gases; middle-weight distillates such as kerosene and diesel oils; still heavier products such as fuel oils; and, finally, the bitumen used in roads and construction, lubricants, and detergents. These basic stock categories correspond to the molecular sizes of the hydrocarbons. Each category has a different boiling point.

In the first stage of the fractional distillation process, the crude oil is pumped from storage into the atmospheric crude distillation unit and heated; at around 200 to 250°F (93 to 121°C), the crude oil enters the desalter. (If the salt were not removed, it would form blockages in the tubes of the furnace.) The desalted crude is then pumped into a trayed distillation tower. The oil vapor rises in the tower and is condensed selectively. The portion that vaporizes in each predetermined temperature fraction is collected and removed from the tower through distillate side streams. A typical mix of side-stream products includes (from top to bottom) kerosene, light gas oil, and heavy gas oil.

In addition to the various fractions condensed and collected, there is a portion that does not condense. This portion is known as the overhead distillate, and it consists of full-range naphtha. The unvaporized residue remaining at the bottom of the tower is fuel oil. Both of these by-products usually need to be processed further. The overhead distillate from the first process is then sent through the debutanizer, which separates butanes and other lighter materials into another overhead distillate.

This distillate is then sent through the depropanizer, which separates a butane stream from the propane and other lighter materials. The remaining substances are sent through the de-ethanizer, which collects a propane stream. The debutanized naphtha (from the debutanizer) is then further divided into a light and heavy naphtha through a naphtha splitter.

By the end of the process, the products normally obtained from crude oil include volatile products (propane, butane, and light naphtha), light distillates (gasolines, heavy naphtha, kerosene, and jet fuels), middle distillates (automobile diesel, heating oils, and gas oils), fuel oils (marine diesel and bunker fuels), lubricating oils, waxes, and bitumen (asphalt and coke).

Chemical refining processes

After the crude oil is distilled, other refinery processes are used to purify the product or transform one type of petroleum to a more useful one. Two processes are used to do this last job: first, cracking, a technique that uses high temperatures to break larger molecules into smaller ones; second, polymerization, which builds up larger molecules from smaller ones (see POLYMERS).

These processes are collectively known as catalytic reforming, and they enable the restructuring of one type of hydrocarbon molecule into another. Catalytic reforming was first developed around 1940 to convert naphtha into high-octane gasoline. The core technique involves using a series of chemical reactions. The feedstock (raw material) is first pretreated with hydrogen to remove sulfur and other nonhydrogen substances; then reactions are carried out over a platinum catalyst in the presence of hydrogen.

Both cracking and polymerization are used to increase the quantity of gasoline produced. Only around 40 percent of crude oil, for example, is made up of gasoline. Catalytic reforming processes are used to convert another 30 percent to gasoline.

To a refiner, what constitutes the best crude varies. However, in general, the lighter crudes are more highly valued than the heavier ones owing to the premium products produced, such as gasoline.

Removal of sulfur

Another key part of the refining process is the removal of sulfur, a solid nonmetallic element found in both organic and inorganic compounds. If sulfur is not removed, then harmful sulfur oxide emissions are released when the fuel is burned.

Refineries use hydro-desulfurization processes to remove the elements. These processes involve treating the petroleum with hydrogen in the presence of a suitable catalyst; this catalyst releases hydrogen sulfide and low-molecular-weight hydrocarbons containing sulfur (mercaptans). These by-products can then be used to make sulfuric acid.

Another desulfurization technique is the Claus Kiln process, which involves passing a mixture of the incompletely burned gases over an aluminum oxide catalyst.

S. FENNELL

See also: ELECTRICITY; FUNCTIONAL GROUPS; HYDROCARBONS; MINING AND PROSPECTING; NATURAL GAS; PETROCHEMICALS; REFINING; SEDIMENTARY ROCKS.

Further reading:

Gluyas, Jon. G. 2004. *Petroleum Geoscience*. Malden, Mass.: Blackwell Publishing.

Yeomans, Matthew. 2004. *Oil: Anatomy of an Industry*. New York: New Press.

GASOLINE

The impact of oil has probably been greatest in transportation. Readily available gasoline, a chief by-product of oil, has made the private automobile a central feature in modern lives. The spread of car ownership vastly increased mobility. For the first time, people could separate where they lived from where they worked, shopped, and played. A vast infrastructure has been built to support the automobile, and networks of freeways now link cities to other cities and to the sprawling suburbs.

In many countries, gasoline is subject to considerable regulation. Burning gasoline in engines produces harmful pollutants. For this reason, there is a global trend toward "cleaner" fuels and alternative energy sources.

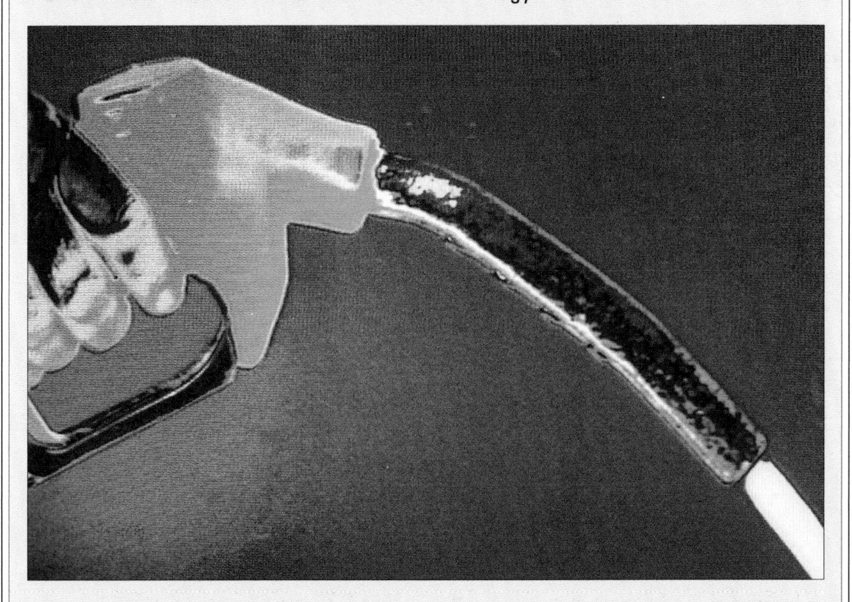

A digitized image of a gasoline pump nozzle: the gasoline appears yellow.

SCIENCE AND SOCIETY

OLBERS'S PARADOX

Olbers's paradox states that the entire sky, day or night, should be as bright as the Sun

German physician and amateur astronomer Heinrich Wilhelm Matthäus Olbers (1758–1840) stated that, assuming the universe is infinite, static, and uniform, the night sky should be as bright as the Sun.

The simple observation that the night sky is dark has important implications for cosmology (see COSMOLOGY). If the universe is infinite, static, and uniform, then a line of sight in any direction should eventually reach a star (see STARS; UNIVERSE). The sky should then appear to be filled with the overlapping disks of stars, and every point in the sky would be as bright as the surface of a typical star. The Sun is a typical star of average size and brightness, so this reasoning suggests the entire sky, day or night, should be as bright as the Sun (see SUN).

These observations were made by many astronomers, including German astronomer Johannes Kepler (1571–1630) in the 17th century and English astronomer Edmond Halley (1656–1742) in the 18th century. The problem is known as Olbers's paradox, named after Heinrich Wilhelm Matthäus Olbers (1758–1840), a German physician and amateur astronomer who wrote about it in 1823.

In Olbers's time a resolution that was widely accepted indicated that interstellar dust would block some of the starlight, so some lines of sight would not encounter stars. This reasoning is invalid, however, because the laws of thermodynamics, which had not then been developed, can be used to prove that interstellar dust that absorbs enough energy from stars eventually would heat to the point that it would begin to glow itself (see THERMODYNAMICS). Thus, even if a line of sight did not end on a star, it would get as far as the hot dust that would shine as brightly as stars.

Invalid assumptions

Astronomers in the 20th century realized that the fundamental assumptions inherent in Olbers's paradox were incorrect: the universe is neither infinitely old nor infinitely large. Because the speed of light is finite, the light from objects that are millions of miles away takes a long time to reach Earth, and these objects are seen as they were in the distant past when they were much younger. If terrestrial observers look sufficiently far, the view would be of a very young universe, even before stars formed (see QUASARS). Furthermore, the universe is finite in size, and there simply are not enough stars in it to guarantee that every line of sight will terminate on a star. Therefore, the conclusion that the sky should be bright is not valid.

The finite extent of the universe is sufficient to resolve the paradox, but there is a flaw in another of the assumptions: the universe is not static. It is expanding, and the recession speed of galaxies (and the stars they contain) increases with their separation (see BIG BANG THEORY; GALAXIES; INFLATIONARY UNIVERSE). This phenomenon has two consequences for Olbers's paradox. Light reaching Earth from a receding galaxy at any one time has farther to travel than the light emitted earlier. Thus, the rate at which light from these distant sources arrives at Earth is decreasing. In addition, the light is redshifted, so the frequency and therefore the energy reaching Earth is reduced (see DOPPLER EFFECT; REDSHIFT). These two effects lower the contributions of distant stars to the brightness of the sky below what would be observed if Olbers's assumption that the universe is unchanging were true. Although the expansion of the universe by itself turns out to be insufficient to produce a dark sky, it does contribute to the explanation of how dark the sky is.

The observed darkness of the night sky constrains cosmological theories, and any theory, such as the extremely successful big bang theory, must resolve Olbers's paradox to be viable.

M. RAYMAN

See also: ASTRONOMY; BIG BANG THEORY; COSMOLOGY; DARK MATTER; INFLATIONARY UNIVERSE; QUASARS; SOLAR SYSTEM; THERMODYNAMICS; UNIVERSE.

Further reading:

Pecker, Jean Claude. 2001. *Understanding the Heavens: Thirty Centuries of Astronomical Ideas.* New York: Springer.

CONNECTIONS

- The **MILKY WAY** is a spiral **GALAXY** consisting of several billion **STARS**, one of which is the **SUN**.
- **REDSHIFTS** of objects within Earth's **GALAXY** are due to the **DOPPLER EFFECT**.

OORT CLOUD

The Oort cloud is a vast collection of comets orbiting the Sun at distances of up to 100,000 astronomical units

Through much of the 19th century and the first part of the 20th century, astronomers puzzled over the source of comets (see COMETS). These objects were known to have a wide range of orbital characteristics, including periods ranging from just a few years to thousands of years. Some had orbits that remained in the vicinity of the planets, while others passed through the inner solar system and traveled far beyond any of the planets.

To explain the observed orbits of the comets, some astronomers theorized that the comets came from interstellar space and, upon passing through the solar system, were captured by gravitational interactions with the Sun and planets (see SOLAR SYSTEM). Other astronomers believed that the comets were produced in the solar system and that their orbits were perturbed by the planets into the elongated ellipses observed today. However, none of the explanations given exactly reproduced the observed characteristics of the population of comets.

Oort's theory

In 1950 Dutch astronomer Jan Hendrik Oort (1900–1992) proposed that the source of long-period comets (those with orbital periods greater than 200 years) was a vast spherical cloud of comets orbiting the Sun at distances from 100,000 to 300,000 astronomical units (one astronomical unit is the mean distance between the Sun and Earth, or about 93 million miles [150 million km]). Oort suggested that every 100,000 to 200,000 years, the gravity from a passing star would send some of the comets back to the inner solar system (see GRAVITY). He estimated that, to match the observed number of new comets, the population of comets in the cloud must be in the order of 190 billion.

Oort's theory rapidly received widespread acceptance. Although the original theory did not explain why there should be so many cometlike objects so far from the Sun, it did provide a reasonable explanation as to the source of long-period comets. While the principal experimental constraints on the theories that describe the cloud continue to rest on the orbital characteristics of comets observed when they are within a few astronomical units of the Sun, astronomers have made highly detailed simulations of cometary orbits and combined them with newer theories of the solar system.

In this expanded version of Oort's theory, as the presolar nebula collapsed to form the Sun and the planets 4.6 billion years ago (see ASTROPHYSICS; NEBULAS; STARS), comets formed in its outer regions, near the orbits of Uranus and Neptune (see NEPTUNE; URANUS). Through gravitational interactions with the planets, the comet orbits were enlarged. Later perturbing effects from other objects in Earth's neighborhood, the Milky Way, led to even larger orbits (see GALAXIES; MILKY WAY). Some comets were ejected completely from the solar system, but others remained gravitationally bound at distances of up to 200,000 astronomical units, or about 3 light-years. (One light-year is the distance traveled by light through a vacuum in one year, or about 5.9 trillion miles [9.5 trillion km].) Additional perturbations from the passage of large concentrations of hydrogen gas plus other molecules (called giant molecular clouds) or, less frequently, from randomly passing stars sometimes shifts the orbits. This shifting causes some comets to fall back into the inner solar system, where they can be observed from Earth.

M. RAYMAN

The artist's impression above shows the Sun in the center of the Oort cloud (ring of small blue objects).

See also: ASTRONOMY; ASTROPHYSICS; BIG BANG THEORY; COSMOLOGY; GALAXIES; METEORS AND METEORITES; NEBULAS; NEPTUNE; PLANET X; PLUTO; UNIVERSE; URANUS.

Further reading:

Seeds, Michael A. *The Solar System*. 2005. Belmont, Calif.: Thomson Brooks/Cole.

CONNECTIONS

- The **GRAVITY** of passing **STARS** and other objects in the **MILKY WAY** causes some **COMETS** to fall back into the inner **SOLAR SYSTEM**.
- Comets move in highly elliptical orbits around the **SUN**.

OPTICAL MINERALOGY

Optical mineralogy is the study of minerals through their interaction with light

A thin slice of mineral cinnabar (mercury sulfate) viewed in unpolarized light through a petrographic microscope. The image is magnified 100 times.

When analyzing the composition of rocks in the field, geologists pick out the component minerals using a hand lens. With this most essential tool of the trade, they perform an audit on a variety of visual clues such as habit (crystal structure), color, luster, and cleavage (the way the stone breaks; see MINERALS AND MINERALOGY). In addition to the lens, experienced geologists also carry a range of equipment for making more tests. Hammers are good for observing the way minerals fracture; fingernails, penknives, copper coins, and small pieces of glass are needed to test comparative hardness (see MOHS' SCALE); a small unglazed tile works best for streak tests (observing the color of a powdered sample), and some enthusiasts even go as far as to try the taste test. However, these various diagnostic tests work best when minerals are present in large grains. As the grain size diminishes, so does the confidence with which accurate identifications can be made.

While expertize with a hand lens could, in one sense, be counted as a form of optical mineralogy, this field belongs exclusively to the examination of rocks and minerals in ultrathin slices using microscopes. For this examination geologists must get their rock samples back to the lab. Thin slices, or sections, are prepared by gluing a slice of rock, cut with a diamond saw, to a glass microscope slide. The slice is then ground down until it is a mere 0.03 mm (30 microns; µm) thick and then topped with a cover glass. At this thinness, most minerals are translucent and can be studied under a microscope. As well as providing a vastly superior method for mineral identification, thin-section optical mineralogy also enhances the potential for detailed study of minerals, their growth and subsequent modification, and ultimately how rocks are formed (see ROCKS).

CONNECTIONS

- The properties of **LIGHT**, such as **REFLECTION AND REFRACTION** and **POLARIZATION**, make it useful for investigating **CRYSTALS**.
- The **GEOCHEMISTRY** of **IGNEOUS ROCKS** and **METAMORPHIC ROCKS** is determined by the assemblage of their **MINERALS**.

CORE FACTS

- Thin sections are slices of rock are mounted onto microscopic slides. Thin sections are cut so finely that they become translucent, allowing them to be studied with transmitted light using a petrographic microscope.
- Minerals can be classed as optically isotropic or anisotropic. Isotropic minerals are amorphous solids or belong to the cubic crystal system and appear black under crossed polars.
- Anisotropic minerals belong to the tetragonal, orthorhombic, monoclinic, triclinic, hexagonal, and trigonal crystal systems. They split light into two mutually perpendicular beams, an effect known as double refraction.
- By using polarizing filters, fixed at 90 degrees, the properties of anisotropic minerals can be investigated.

Light investigations

In microscopy, the imaging method chosen to investigate a sample is called the probe. The probe can be light in the visual spectrum in the case of optical microscopy, X rays as in X-ray crystallography, or electrons in electron microscopy (see ELECTRON AND POSITRONS; IONS AND RADICALS; LIGHT; X RAYS). As a general rule the spatial resolution of the probe is the minimum size object that can be resolved with that imaging method. The X rays used in X-ray crystallography can be used to give information about crystal structure because their wavelengths (0.1–0.2 nanometer; nm) are of the same order of magnitude as the size of atoms themselves. Electrons, similarly, have energies that correspond to wavelengths of 10 to 100 nm, a size that allows objects from 10 to 100 nm (0.1–0.01 microns) to be imaged.

Visible light, on the other hand, has wavelengths that range from 390 nm and 710 nm. Since the unit cells of most crystals are many hundreds or thousands of times smaller—typically from 0.2 to 5 nm—it is clear that visible light is not affected by the intricacies of crystal structure (see CRYSTALS AND CRYSTALLOGRAPHY). Instead, it is the bulk properties of crystal symmetry that affect light. The influence of huge blocks of unit cells, ranks of repeating components thousands across, is what is seen through a microscope, along with the defects that affect crystals at this scale. Although the perfect microscopes would, in theory, have a resolution limit of around 500 nm, in practice it is more like 1000 nm (1 micron).

Of light and crystals

The speed of light in a vacuum is a universal constant, but when light travels through matter, its speed is reduced. Visible light is simply a certain bandwidth of the electromagnetic spectrum to which the eye's optic nerves are sensitive (see ELECTROMAGNETIC SPECTRUM). Like all electromagnetic radiation, it propagates as a series of electrical and magnetic oscillations. These oscillations vibrate in planes perpendicular to the direction the radiation travels. Since all matter contains electrical fields associated with its constituent atoms, passing light rays interact with these fields, causing them to slow down (see ATOMS).

The speed of light in a substance depends on the density and strength of the electric field caused by its atoms. The denser a substance, the more atoms it has per unit volume. The closer the atoms are together, the greater the strength of the electric fields and the greater the retardation of light. This effect is well demonstrated by two alumino-silicate minerals, sillimanite and kyanite. Both have the same composition (Al_2SiO_4), but kyanite is slightly denser at 0.8 ounces per cubic inch (3.53 g/cm^3) compared with sillimanite's 0.73 ounces per cubic inch (3.23 g/cm^3). Accordingly, light travels more slowly through kyanite, at approximately 109,000 miles per second (175,000 km/s) than through sillimanite (112,500 miles per second; 181,000 km/s).

Similarly, large atoms with high electron densities have a concomitant greater effect upon the speed of light traveling through them. Periclase (manganese oxide; MgO) and wüstite (iron oxide; FeO) are very close in density, but iron with 26 electrons has a greater electron density than magnesium with its 12 electrons. With more electrons per unit volume, light travels though wüstite at approximately 80,000 miles per second (129,000 km/s), compared with 107,500 miles per second (173,000 km/s) for periclase.

PETROLOGY

The study of minerals in thin section has led to the development of the sciences of petrology and petrogenesis. Petrology is a blanket term for all aspects of the study of rocks, especially their mineralogy, textures, structures, alterations, and associations. The study of their origins, development, reactions, and modifications is the province of petrogenesis. Since igneous rocks are almost always matrices of interlocking crystals and metamorphic rocks often have significant amounts of crystal regrowth, mineralogy has a lot to say about how these rocks are formed. For example, rocks composed of large crystals have cooled slowly from liquid magma, perhaps in large underground batholiths of molten rock, giving them the time to grow fully formed shapes. Small crystals conversely have formed quickly. Those rocks that have large crystals surrounded by a finer-grained matrix tell a story of slow-growing crystals suspended in hot magma suddenly ejected. The mineral assemblages that occur in rocks also provide clues to the chemical makeup and conditions prevailing as the rock formed. Changing chemistry is often reflected in artifacts, such as inclusions and zoning.

A CLOSER LOOK

This change in the speed of light as it travels from one substance into another causes a light ray to bend, or refract. The higher the frequency of light, the more it refracts (see REFLECTION AND REFRACTION). Refraction is what causes white light to split into its component colors when it passes through a prism—the phenomenon famously explained by Isaac Newton (1642–1727). As a direct consequence of their internal atomic arrangement, different translucent materials display the ability to refract light to a greater or lesser extent. This property is quantifiable, and its measurement is called the refractive index. The refractive index of a material is given as

$$\text{refractive index} = \frac{\text{speed of light in a vacuum}}{\text{speed of light in material}}$$

Since the speed of light is greatest in a vacuum, the value of the refractive index is always more than one.

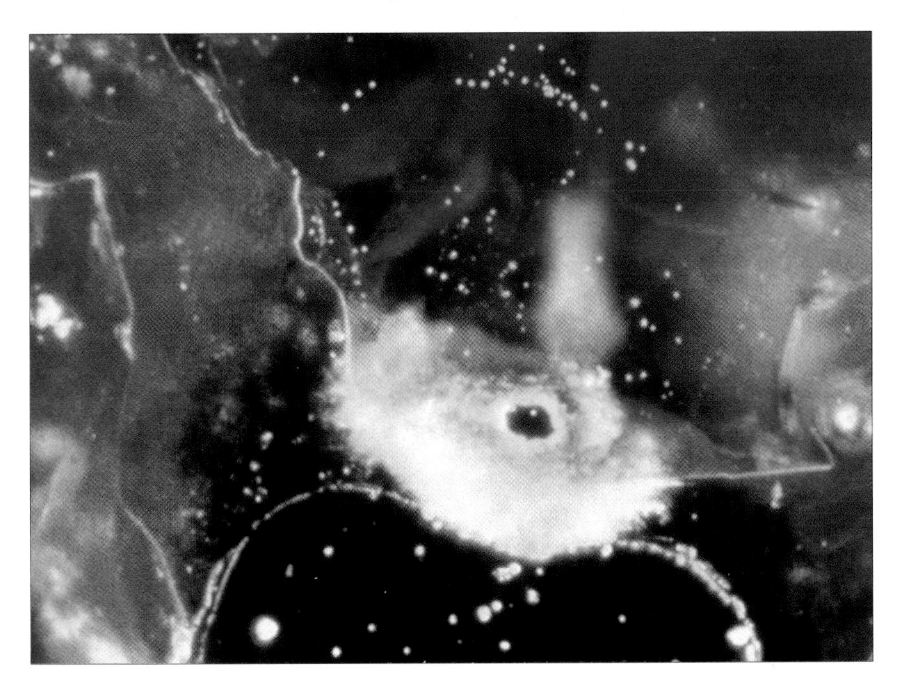

Vitrified (glassy) alumina viewed through a light microscope. The image is magnified 500 times.

HENRY CLIFTON SORBY

Henry Clifton Sorby (1826–1908) is the father of optical mineralogy. The Englishman is one of history's most foremost scientists and one of the pioneers of petrography (the study of rocks in hand specimen and thin section), but his work is largely unknown, and his contribution remained unrecognized until it was chanced upon and publicized by the German microscopist Ferdinand Zirkel. Self-financed, Sorby (pictured below) took up the study of minerals in thin section, observing them both under normal and plane-polarized light. He applied himself to solving the problem of how slates can have cleavage planes that bear no relation to their bedding planes—so-called slaty cleavage. He proved in 1853 that mechanical pressure had realigned the clay minerals, a breakthrough for his microscope technique, which many scorned (see CLAY).

"In those early days people laughed at me. They quoted Saussure who had said that it was not a proper thing to examine mountains with microscopes, and ridiculed my action in every way. Most luckily, I took no notice of them."

In 1857 Sorby was elected a fellow of the Royal Society and in 1869 awarded the Wollaston medal by the Geological Society of London. During his lifetime he held the presidency of the Royal Microscopical Society, the Mineralogy Society, and the Geological Society.

DISCOVERERS

Isotropic and anisotropic minerals

The ordered internal arrangement of atoms in crystals fit one of seven crystal systems. These systems describe how the unit cells piece together and govern the bulk symmetry of a crystal. They are defined by the crystallographic axes, a construct that is used to categorize crystal geometries. Minerals of the cubic crystal system, such as table salt and diamonds crystals, have crystallographic axes of equal length, each at 90 degrees to the other. From the point of view of an incoming light ray, the crystal looks the same in every direction. The speed of light will be the same whichever direction the light waves travel through the crystal, and there is one value for the refractive index of the mineral. The same is true for amorphous materials, such as glass, which have no true crystal structure. Hence, minerals in the cubic system and amorphous materials are called isotropic, having the same properties in all directions.

However, some minerals can have more than one refractive index. If a mineral has more than one unique crystal axis, the crystal is denser (or has a stronger electric field) in one direction than another. An incoming light ray is split into two mutually perpendicular beams, one traveling slower than the other. Minerals that display this property are call anisotropic, literally "not isotropic." This splitting of light results—an effect called double refraction—can be seen very clearly in the mineral calcite. Looking at an object through a crystal of calcite produces a double image because the light from the object reaches the eyes by two separate paths. Anisotropic minerals in the tetragonal, hexagonal, and trigonal crystal systems have two unique crystallographic axes and therefore two refractive indexes. Those in the orthorhombic, monoclinic, and triclinic crystal systems have three unique crystallographic axes and hence three refractive indices.

These two, linked discoveries—that minerals in the same crystal system affect light in the same way and that anisotropic minerals spit light rays—allowed pioneers, such as Henry Clifton Sorby, to realize the potential for studying minerals under the microscope. The final innovation was the development of the petrographic, or petrological, microscope (see box on page 1104).

Under the microscope

The different optical properties of minerals produce effects under transmitted light that are used as diagnostic markers under the microscope. To identify a mineral, the crystal system is determined along with other attributes, such as color, cleavage, relative refractive index, as well as other common traits, such as twinning or alteration. The central innovations of the petrographic microscope are the polarizing filters—the polarizer and the analyzer—that allow the user to view thin sections under plane-polarized light (see box at right). Under crossed polars (with both the polarizer and the analyzer engaged) all isotropic minerals—minerals in the cubic crystal system and amorphous minerals—immediately appear opaque or black, and thus, the mineralogist can distinguish between them.

Principal mineral features

• **Relief:** Under the microscope some minerals stand out in sharp relief, their edges crystal clear and their surface textures, such as cleavage, well defined, while others appear flat and almost featureless. This property, known as relief, depends on the mineral's refractive index relative to that of the mounting medium (the glue used to mount the rock onto the microscope slide). If the difference between the refractive indexes is large, the mineral shows up clearly, and its relief is described as high. If, however, the refractive index of the mineral is close to that of the mounting medium, then the mineral will display a low relief. Relief is a useful diagnostic test for distinguishing between minerals that are found in igneous rocks. Ferromagnesian minerals, made of iron and magnesium, such as olivines, show high relief, whereas most silicates and feldspars show low relief (see MINERALS AND MINERALOGY).

• **Cleavage:** Most minerals contain characteristic planes of weakness in their atomic lattice along which they break when subjected to stress. These planes can be seen under the microscope as lines and cracks. Often a mineral contains more than one cleavage plane. In this case, care needs to be taken to determine the angle at which the crystal presents itself. Amphibole minerals display prismatic cleavage, with both cleavage planes at 124 degrees to each other and parallel to the length of the crystal. If an amphibole is seen in thin section cut lengthways down the crystal, it will display only one cleavage running parallel to the long axis, the so called basal cleavage. An end-on amphibole crystal must be found, which shows both cleavage planes at 124 degrees to each other, to make a correct identification.

• **Color and pleochroism:** The color of any given mineral in thin section is due to the selective absorption of certain wavelengths of light from the continuous spectrum of white light shone through the slide. Mineral colors are often thin, pastel colors, such as the light pinks displayed by augite. However, some minerals, notably biotite with its brown and green shades, show stronger colors. Biotite also displays a phenomenon known as pleochroism, where the colour of the crystal in plane-polarized light changes as it is rotated on the microscope stage. This color change, in shades of brown in biotite, is due to the anisotropy of noncubic crystals. Just as the refractive index is different along each unique crystallographic axis, so light absorption can vary along different vibration directions.

• **Extinction and extinction angle:** Analysis of a mineral's extinction characteristics is done with the analyzer engaged on the petrographic microscope. The polarizing filters are set at right angles to each other; therefore, if a mineral crystal does not change the vibration direction of transmitted light passing through it, no light will reach the eyepiece, and the crystal will be dark, or "in extinction."

When either of the crystal's vibration directions is aligned with the privileged vibration direction of the microscope's polarizer, light is transmitted through the crystal only along that vibration direction; thus, when it reaches the analyzer, no component of the light will be allowed through.

However, if the crystal is not in its extinction position, light will travel through it along both vibration directions. When the light encounters the analyzer, some components of it will be aligned with the filter and will pass through to the eyepiece.

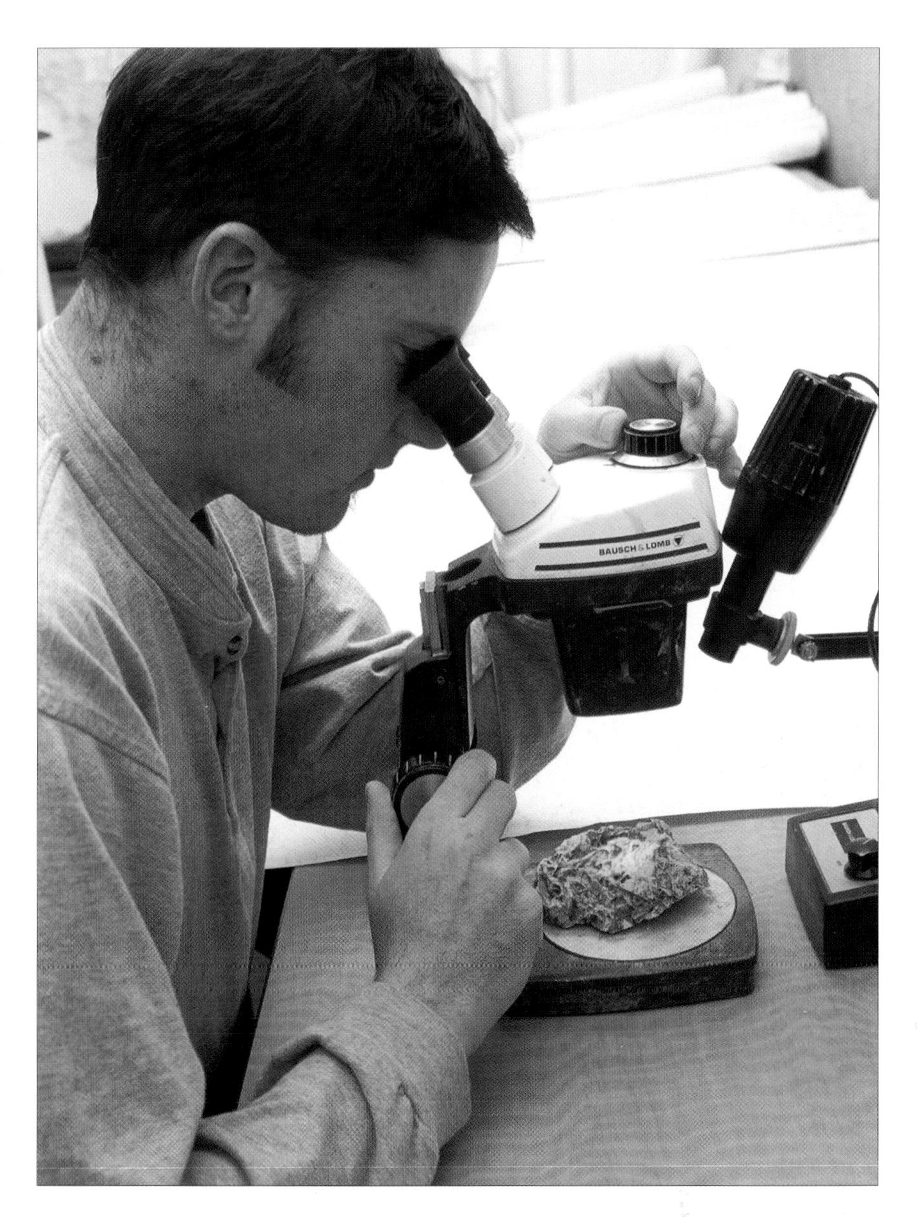

Looking at rocks through a lens is the first step to figuring out what minerals are present.

POLARIZED LIGHT

Light from the Sun (and almost every other light source) is unpolarized (see OPTICS). Thus, electromagnetic oscillations of the individual light waves that make up a ray of light vibrate in every plane perpendicular to the direction of propagation. A ray becomes plane polarized when all the waves vibrate in the same plane (see POLARIZATION). Partial polarization occurs in nature when light reflects off a surface such as water, but polarizing filters, such as those used in Polaroid shades, polarize light totally. These filters, acting as a gate, allow only the waves that vibrate in the chosen plane through.

Plane-polarized light is a very powerful tool for observing anisotropic minerals in thin section. Because noncubic minerals split light into two mutually perpendicular beams with different speeds (refractive indices), both refractive indexes of the mineral can be examined one after the other. Using the rotating stage, first one and then the other principal vibration directions of the mineral are aligned with the privileged vibration direction of the polarizer of the microscope. If the difference between the fast and slow vibration directions of the mineral is large, as it is in carbonates, then crystals of the mineral can be seen to twinkle as the stage is rotated under plane-polarized light.

A CLOSER LOOK

THE PETROGRAPHIC MICROSCOPE

The petrographic, or petrological, microscope uses transmitted light to study thin sections of rocks. In addition to the normal apparatus of a microscope and multiple objective lenses, it also has a rotating stage and two polarizing filters set at 90 degrees to each other. The polarizer plane-polarizes the incoming light in an east–west direction. The analyzer, which can be engaged and disengaged, polarizes light in a north–south direction. The petrographic microscope also features an auxiliary lens in the microscope tube, called a Bertrand lens, which is used for viewing interference figures.

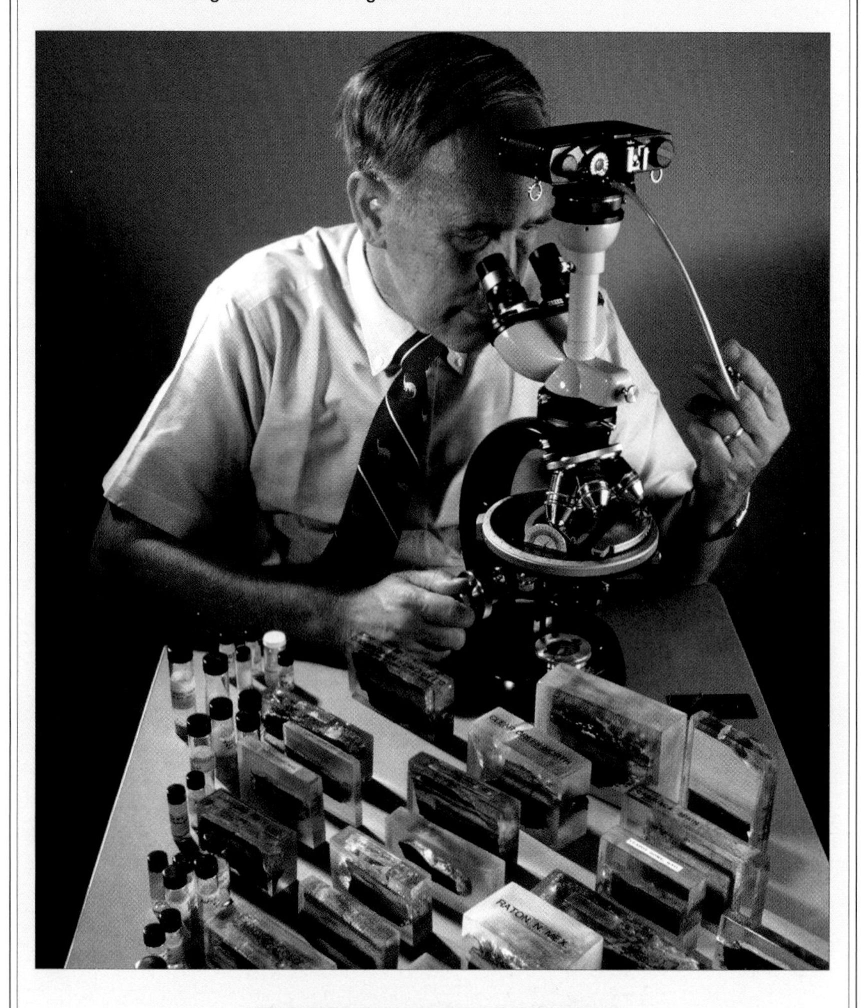

A CLOSER LOOK

Every mineral has four positions, all at 90 degrees to each other, where it will be in extinction. The angle that these positions make with a crystal's cleavage planes is useful for identifying minerals. Straight extinction is when the cleavage is parallel to the eyepiece crosshairs at extinction. Inclined extinction is where the cleavage is at an angle to the crosshairs on extinction. The angle between the cleavage and the crosshairs is the extinction angle. Minerals of the tetragonal, hexagonal, trigonal, and orthorhombic crystal systems show straight extinction, those of the triclinic system show inclined extinction, and those of the monoclinic system generally show inclined extinction but sometimes show straight extinction.

• **Birefringence:** Under plane-polarized light, light travels along both of a crystal's vibration directions when it is in between the extinction positions. If the analyzer is engaged, light from both vibration directions is recombined into a single plane. Because the rays have traveled through the crystal at different speeds, when they recombine, they will be out of phase and will interfere with each other (see LIGHT). The result is a colored ray. Because of the phase difference between the two rays, some wavelengths will interfere constructively with each other, while others interfere destructively. The outcome is that certain colors become reinforced, and others are eliminated.

Birefringence is the numerical difference between the fast and the slow refractive indices in a mineral. It produces a diagnostic maximum in color at the point where the interference between the two light rays is at its greatest. The Michel-Levy color chart enables the birefringence to be read off by matching the maximum color to the thickness of the thin section.

Other diagnostic features

Minerals all display a variety of unique characteristics that the mineralogist becomes expert at spotting. Slabs of striped lamellar twins side-by-side in the same crystal, for example, are a dead give away of plagioclase feldspar. The relatively common occurrence of twinning in crystals is due to the direction of growth changing as the crystal was forming. Changes in a crystal's composition as it forms result in "zoning." These changes often cause alterations of the crystal's refractive index, and give rise to visual effects such as concentric or prismatic banding.

Other minerals are prone to alteration or inclusions. Inclusions occur while the crystal was growing and result in one mineral being engulfed by another with the original crystal or many small crystals becoming trapped inside the latecomer. Alterations occur after the crystal is formed and are compositional changes that affect a mineral's optical properties. Biotite, for example, can alter to chlorite, which shows green rather than brown pleochroism. Small inclusions of radioactive atoms in a crystal lattice can be seen in thin section as black spots or holes where radiation has damaged the crystal lattice.

By extensive practice and experience, the mineralogist builds up a diagnostic toolbox. He can perform a checklist of tests on every type of crystal in thin section of a rock to identify the component minerals. Being able to determine the mineral assemblage of a rock enables it to be classified and the conditions under which it formed to be understood.

D. GREEN

See also: ATOMS; COLOR; DENSITY; ELECTRONS AND POSITRONS. GEOLOGY; MATERIALS SCIENCE; METAMORPHIC ROCKS; MOHS' SCALE; NEWTON, ISAAC; OPTICS; PIEZOELECTRICITY; ROCKS.

Further reading:

Hibbard, M. J. 2002. *Mineralogy: A Geologist's Point of View*. Boston: McGraw-Hill.

Nesse, William D. 2004. *Introduction to Optical Mineralogy*. New York: Oxford University Press.

OPTICS

Optics is the study of the creation, propagation, and detection of light

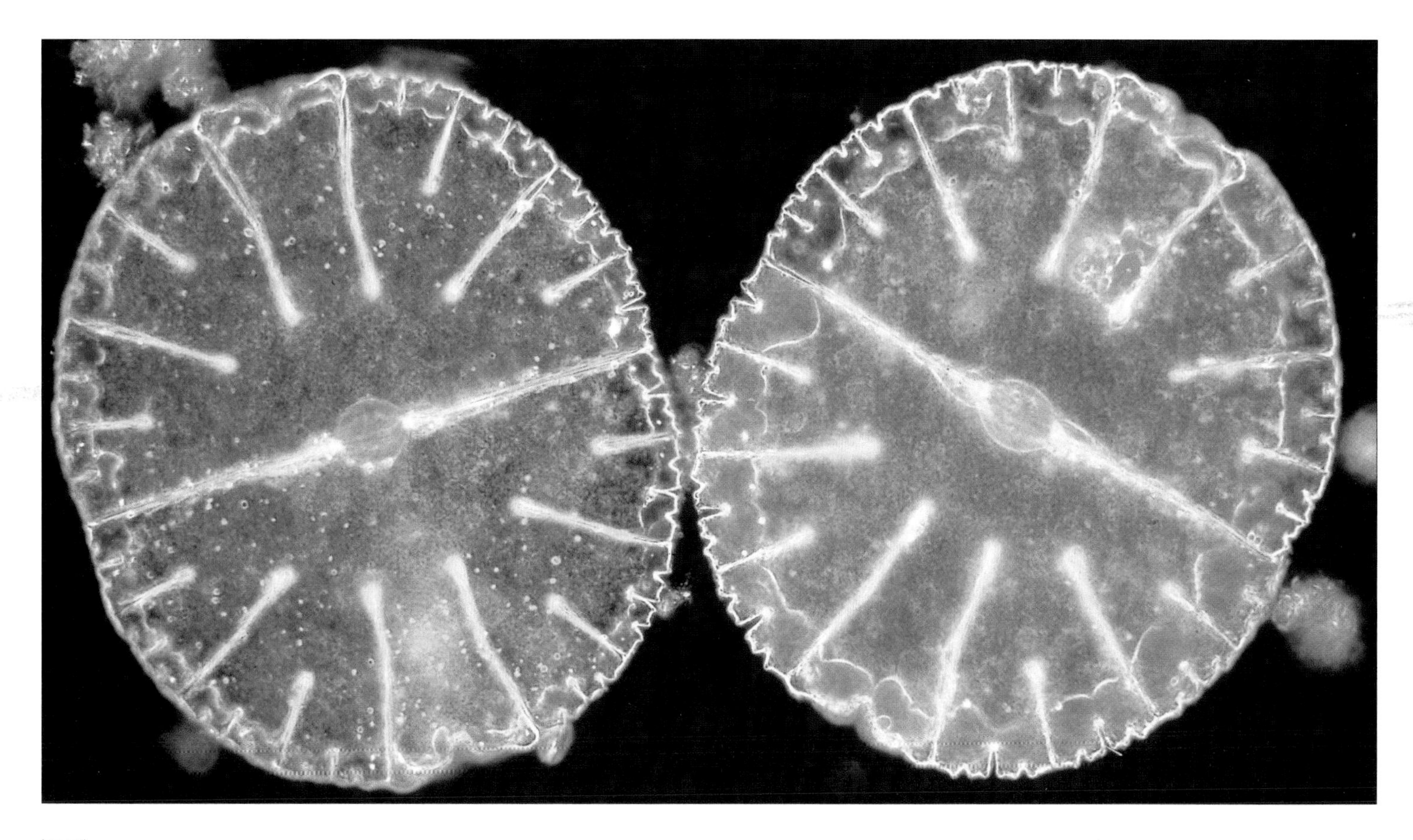

This light micrograph shows a species of desmid, a type of green alga.

The study of light, or optics, has enabled humans to extend their view of the world. Microscopes allow scientists to see details as small as the nucleus of a single living cell. At the other extreme, the Hubble Space Telescope, orbiting Earth at 381 miles (614 km) above its surface, produces images of galaxies at the very edge of the known universe (see SPACE EXPLORATION). Studying light has also led to the discovery of the most fundamental theories of modern physics, such as quantum theory and relativity (see QUANTUM THEORY; RELATIVITY). The modern quantum theory of light took centuries to develop. Light is subtle, sometimes showing both wavelike and particle-like behavior. For this reason, it is a good theme for a case study showing how science develops.

Although the ancient Greeks formulated the law of reflection and recognized the phenomenon of refraction, they could not agree on the nature of light and vision. Plato (428–348 B.C.E.), a student of Socrates (470–399 B.C.E.), believed that the eye sent out beams that cause sight as soon as they meet an object (see REFLECTION AND REFRACTION). Euclid of Alexandria (ca. 300 B.C.E.) had a considerable understanding of geometric optics, but even he held this same faulty view about light and vision.

Euclid is most famous for writing *Elements*, 13 books on geometry that contain much of the geometry studied in today's high school mathematics courses. He also contributed to the study of light; he described the law of reflection and how light travels in straight lines. Combining his knowledge of geometry and the law of reflection, Euclid deduced the focusing properties of a spherical mirror and reported that the intense light rays produced when a concave mirror is pointed toward the Sun would set wood alight.

Aristotle (384–322 B.C.E.), Plato's student, disagreed with his teacher and argued that vision occurred because of something coming from the object and entering the eye. However, Aristotle argued that light itself does not travel. He taught that light was a quality that a medium acquired all at once. For the following 2,000 years, this view of the nature of light was held by many of those who studied optics.

During the Middle Ages in Europe, science continued to flourish in Arabia. One of the foremost of the Arab scholars was Abu 'Ali al-Hasan ibn al-Haytham (or Alhazen; 965–1039). He was the first to understand vision and to describe correctly how the

CORE FACTS

- Optics is the study of the properties of light.
- Light can be modeled as an electromagnetic wave or as packets (quanta) of energy called photons.
- The study of light has led to the most fundamental theories of modern physics, including quantum theory and the theory of relativity.
- Optical instruments include the microscope and the telescope.

CONNECTIONS

- **HOLOGRAPHY** is a technique for producing full three-dimensional images of people and objects.
- Light signals can be transmitted almost perfectly over distances of many miles using **FIBER OPTICS**.
- Optical techniques can be used to produce images using waves such as **ULTRASOUND** and **INFRARED RADIATION**.

When light enters the eye, the cornea and lens cast an image on the retina. This image falls in front of the retina in someone with myopia (nearsightedness), and a concave lens is needed to correct sight. When someone is hyperopic (farsighted), the image falls behind the retina, and a convex lens is required to correct sight.

cornea and lens cast an image on the retina when light enters the eye. He also repeated Ptolemy's experiments on refraction and showed that the angles of incidence and refraction are not proportional. His *Opticae Thesaurus* was translated in the late 13th century and became one of the principal optics texts of that time.

Near the end of the Middle Ages, English philosopher and scientist Roger Bacon (1220–1292) made valuable contributions to the study of lenses. He is sometimes credited with inventing the refracting telescope, but although Bacon suspected that two lenses could be used to make such a device, he apparently never constructed one.

The microscope and telescope

The dawn of the 1600s saw an explosive increase in the number of optical instruments. The microscope and the telescope were invented nearly simultaneously, each in several countries throughout Europe. Claims were made by Zacherarius Jensen of Holland that he had invented the microscope in 1590. George Huefnagel of Frankfurt, Germany, made a similar claim and published drawings of insects reportedly made with a microscope in 1592.

No one is sure who invented the first refracting telescope. What is certain, however, is that Hans Lippershey (1570–1619), an eyeglass maker in Middleburg, Holland, applied for a patent on the device on October 2, 1608. It was denied because a large number of people had already claimed to have invented the refracting telescope, including his countryman Jensen. In any event, it is clear that by 1609 a knowledge of telescopes had spread throughout the major cities of Europe and, in particular, to Italian astronomer and mathematician Galileo Galilei (1564–1642; see GALILEI, GALILEO). Galileo discovered craters on the Moon (see MOON) and the four brightest moons of Jupiter (see JUPITER) and found

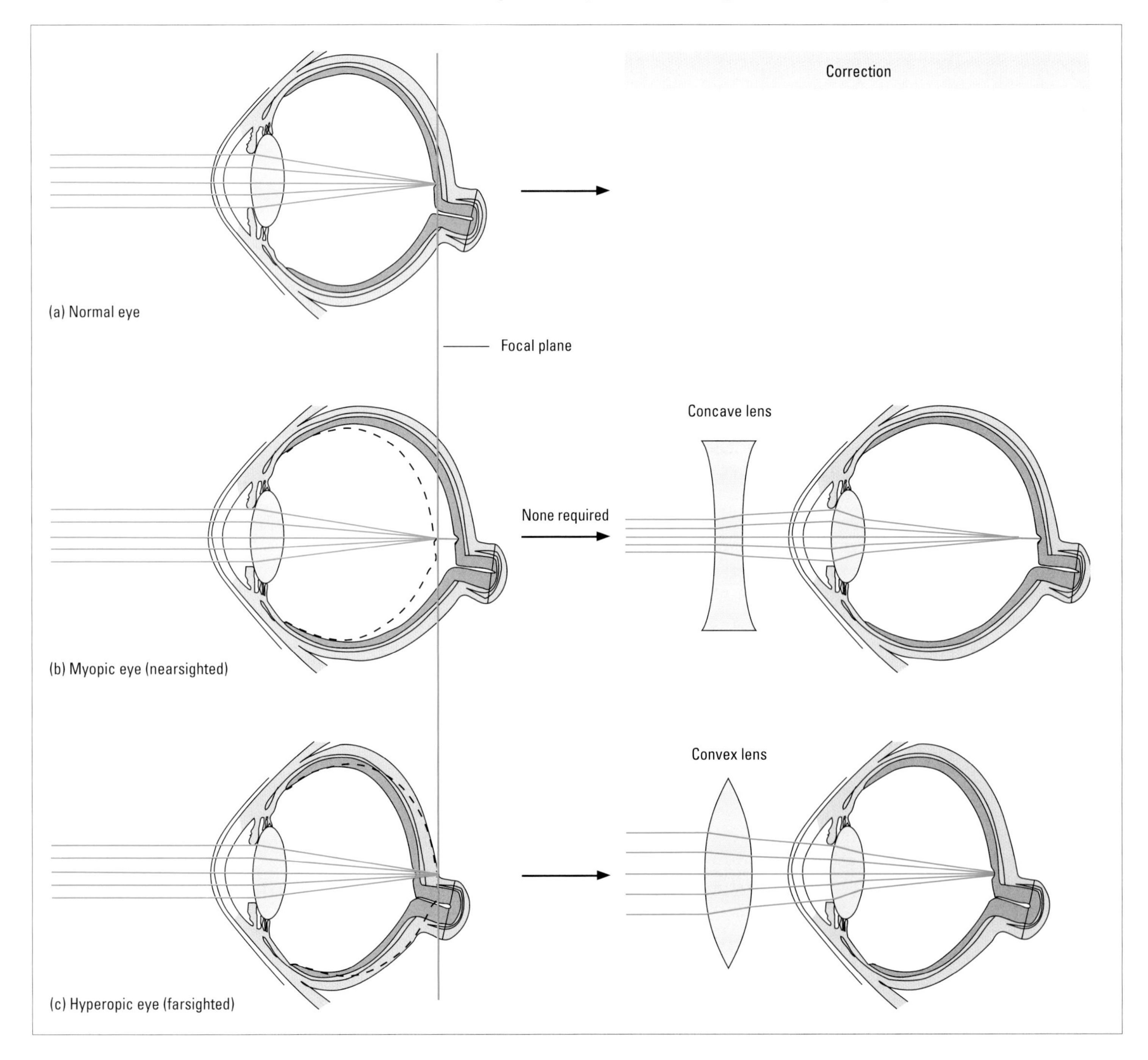

that the Milky Way is composed of countless billions of individual stars (see MILKY WAY). Humankind's view of the universe was altered forever.

In an effort to improve the quality of the image, German astronomer Johannes Kepler (1571–1630) studied how the telescope formed images. In 1611 Kepler published *Dioptrica*, a major contribution to dioptrics, the mathematical study of the refraction of lenses. Kepler used an approximate law of refraction, which stated that for light entering a transparent medium, the ratio of the angle of refraction to the angle of incidence is a constant. Until then, a diverging lens had been used in the telescope eyepiece. Kepler suggested that a convex lens would also work.

The law of refraction

The law of refraction accepted today was discovered by Dutch astronomer and mathematician Willebrord Snell (1580–1626) around 1621. Snell did not publish his result and may not even have realized its importance. His discovery is included in reports published by Dutch mathematician and physicist Christiaan Huygens (1629–1695) and Isaak Voss, who saw Snell's unpublished work. Snell reported his result as a ratio of trigonometric functions. He found that the ratio between the sines of the angles of incidence and refraction was equal to the ratio of the velocities of light in any two materials (see REFLECTION AND REFRACTION).

The modern form of the law of refraction was first published by French mathematician and philosopher René Descartes (1596–1650) in 1637. Descartes claimed to have derived the law using his method of logical deduction. There is some speculation, however, that he may have been aware of Snell's work and actually worked backward from the answer. Although Descartes's proof was greatly flawed, its mathematical nature and the assumption that the speed of light is different in different media became stepping stones for later progress.

Understanding light

Descartes's *Dioptric*, published in 1637, highlighted the contradiction in the theories of light at that time and set the stage for a major transition in the understanding of light. On one hand, there was Descartes who, like Aristotle, Alhazen, Bacon, and others of the "perspectivist" school of thought, argued that light is a kind of luminous pressure transmitted instantaneously by a rigid medium between the source and observer. Descartes argued that there is no need to assume that something material passes from an object to the eyes to produce sight. On the other hand, Descartes's descriptions of reflection and refraction modeled light as if it were made up of projectiles traveling with finite speed and obeying laws similar to the mechanics of particles having mass.

Pierre de Fermat (1601–1665), a French mathematician, was a vocal critic of Descartes's derivation of the law of refraction. He found Descartes's treatment, which involved both an actual and an intended direction of motion, unacceptable. Fermat (and others)

NEWTON AND THE REFLECTING TELESCOPE

One of the troubling properties of early refracting telescopes was chromatic aberration, the tendency of lenses to bring light of different colors to focus at different distances from the lens. The cause of this aberration was not understood and scientists hoped a properly shaped lens would eliminate the effect. English physicist and mathematician Isaac Newton (1642–1727) investigated the problem and discovered that not all rays from the Sun are refracted identically. When white light from the Sun is passed through a circular hole and then through a prism, the light disperses into an oblong rainbow. Newton found that if a single portion of the rainbow is made to pass through another small hole and a second prism, the light does not disperse farther but remains the same color. Furthermore, when the full rainbow is passed through a second prism held upside down, the colors are recombined into a white circular pattern.

Through these experiments, Newton came to realize that different colors of light are bent in differing amounts when they pass through glass. Newton concluded that refracting telescopes would always have chromatic aberration. As a consequence, in 1668 he designed a telescope that used a concave mirror to gather light. The mirror forms a real image of the object, and this image is examined with a magnifying eyepiece.

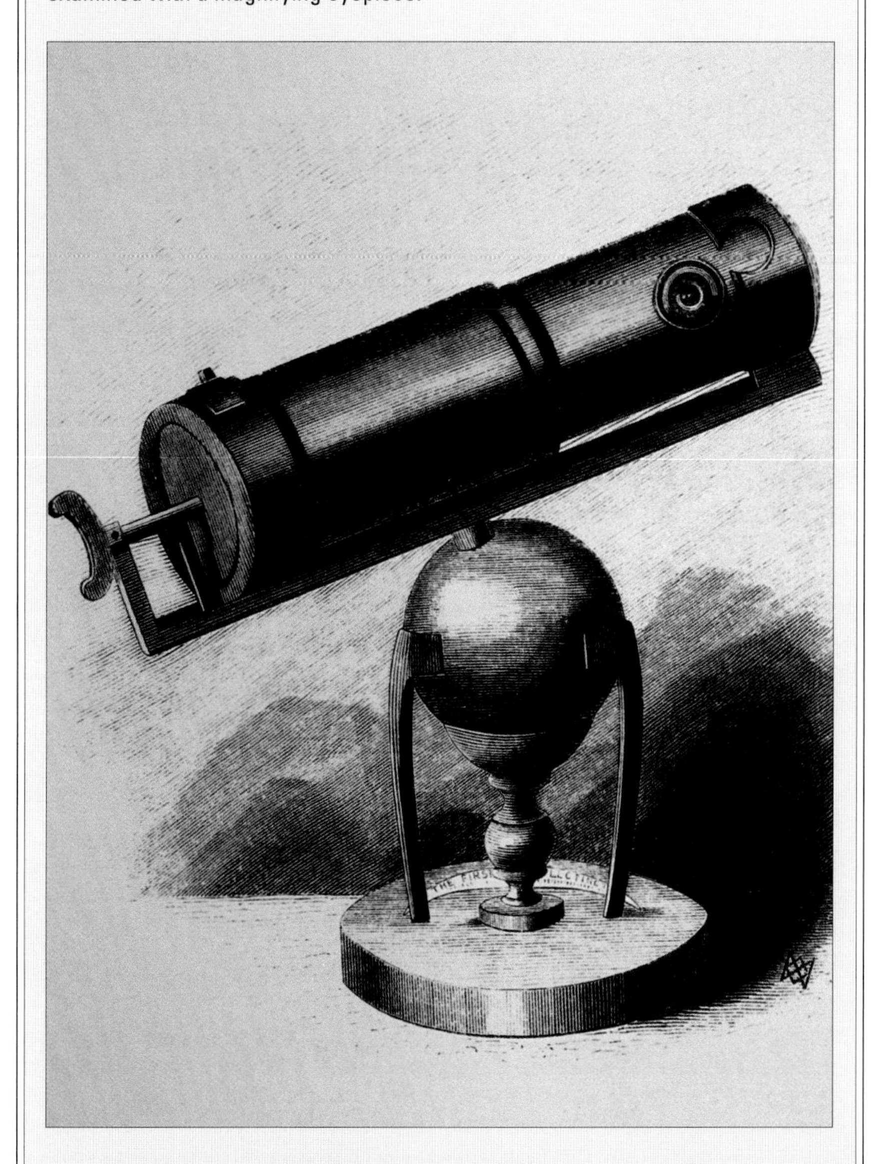

Isaac Newton's reflecting telescope, designed in 1668, used a mirror to gather light and had a magnification of x 38.

HISTORY OF SCIENCE

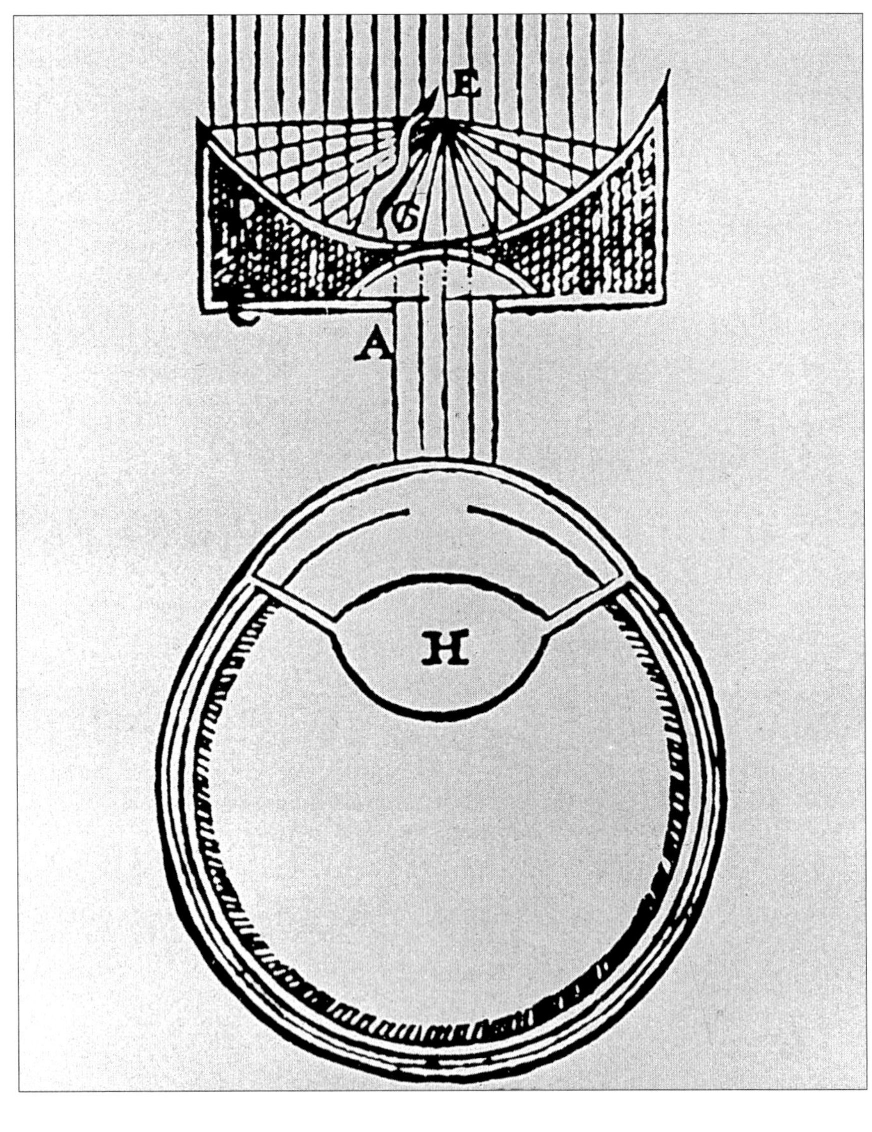

This drawing of a simple microscope comes from Descartes's* Dioptric, *published in 1637. It has a single lens, the specimen is impaled on a spike, and a silvered mirror provides the illumination.

argued that Descartes's assumptions were selected merely to reproduce the observed behavior and did not provide any insight into the behavior of light.

Fermat derived the law of refraction from the principle of least time, now called Fermat's principle. It had been known since the early Greeks that the law of reflection could be derived from the principle that light travels between two points along the shortest path. Fermat's insight was to extend this idea to refraction by stating that light would take the path that resulted in the least travel time.

Waves or particles?

Descartes's derivation of the law of refraction started a debate that lasted until years after his death. The major outcome of the debate was that the long-standing Aristotelian view of the instantaneous propagation of light was abandoned. Instead, light came to be viewed as something that travels through space with finite speed. However, a new controversy arose between those who believed light was a wave and those who believed it was made of particles.

Building on a rough outline of a wave theory written by British scientist Robert Hooke (1635–1703) in 1665, Dutch mathematician and physicist Christian Huygens (1629–1695) became the first important proponent of the wave theory of light. Huygens imagined space to be filled with an elastic medium, called the ether, that transmited vibrations from source to observer, in the same way that air transmits sound waves. In his *Traité de la Lumière* of 1690, Huygens imagined wave fronts propagating outward from the light source, just as water waves propagate outward when a stone is dropped into water. Each point on the wave front becomes a source for additional wavelets that combine to produce the total wave front. To explain refraction, Huygens's wave theory required that light travel more slowly in water and glass than in air. Opponents of the wave theory of light maintained that light is composed of particles that travel through space from source to observer. Chief among proponents of the particle theory was English physicist and mathematician Isaac Newton (1642–1727; see NEWTON, ISAAC). Newton based his objection to the wave theory on the fact that light did not exhibit the well-known tendency of waves to bend around obstacles into the shadow region, a phenomenon known as diffraction. In direct contradiction to the wave model, Newton's particle theory required that light particles travel faster in glass than in air. Such was Newton's reputation that when he declared that light consists of streams of particles, the wave theory was abandoned for nearly a century.

A victory for the wave model

The wave theory was revitalized by British physicist Thomas Young (1773–1829). In a paper presented to

PHOTON MOVEMENT

Quantum electrodynamics (QED) considers the interactions between light and matter. It is based on the idea that charged particles interact by emitting or absorbing photons, particles of light. The photons cannot be observed; one can only calculate the likelihood that a photon will take a particular path from one place to another. Scientists assign a probability amplitude to each path, which can be represented by an arrow. The greater the likelihood that the photon will take the path, the longer the arrow will be. These arrows point in various directions in space. Until the photon is emitted, the arrow rotates at speeds to the order of 10^{14} rotations each second. To calculate the probability that a photon will travel between two points, one must add up the arrows belonging to all the possible paths and chain the arrows together, head to tail. Finally, a single arrow is drawn from the beginning of the first arrow to the end of the last. The length of this final arrow is then squared. The resulting number is the probability that the photon will travel from the first point to the second. Sometimes the arrows representing nearby paths point in opposite directions. Consequently, when all the arrows are added up, they cancel each other out, the result being a smaller or even zero-length arrow. On the other hand, sometimes the arrows add up in the same direction, making it more likely that the photons will follow a straight path, like the one a pool ball might follow on a table.

the Royal Society in 1801, Young used the wave theory of superposition to explain the colors seen in thin films of oil on the surface of water. He later explained the formation of light and dark bands formed when light passes through two adjacent narrow slits. Despite these successes, Young's theories were ridiculed for 20 years. Eventually, support for his theories grew in France when French physicist Augustin-Jean Fresnel (1788–1827) made observations of diffraction patterns around a thin wire (see DIFFRACTION AND INTERFERENCE).

In 1818 Fresnel entered his theory of diffraction into a competition sponsored by the French Academy of Science. One of the judges, French mathematician Simeon D. Poisson (1781–1840), was a strong opponent of the wave theory of light. To prove how ridiculous Fresnel's theory was, Poisson showed that if monochromatic light (light of one wavelength) was used, Fresnel's theory predicted that the shadow of a small circular disk would have a bright spot in the middle. Another of the judges, French physicist François Arago (1786–1853), tried the experiment, and to the surprise of many, the bright spot was actually there. Fresnel's paper won first prize.

By 1825 Young, Fresnel, Arago, and others had worked out a detailed description of the wave theory of light, including the realization that polarization effects require that light be a transverse wave with oscillations perpendicular to the direction of propagation. The final blow to the particle theory came in 1850 when French physicist Jean-Bernard-Léon Foucault (1819–1868) measured the speed of light in water and found it to be less than the speed of light in air, just as predicted by the wave theory.

During this same period, great strides were made in understanding electricity and magnetism. Scottish physicist James Clerk Maxwell (1831–1879) summarized the experimental results of electricity and magnetism in a set of four equations that bear his name (see ELECTROMAGNETISM). On studying these equations, Maxwell found solutions that described transverse electromagnetic waves propagating freely through space. When Maxwell calculated the speed of these waves from numbers measured in independent electric and magnetic experiments, he found the speed to equal the speed of light. The conclusion was unavoidable: light is an electromagnetic wave.

Return of the particle model

At the close of the 19th century, the model of light as a transverse electromagnetic wave was well established. However, there remained two troubling problems that the theory could not explain. One was that of the distribution of light given off by objects when they are heated. British physicists James Jeans (1877–1946) and Lord Rayleigh (John William Strutt; 1842–1919) developed the classical model that gave correct results only for long wavelengths. German physicist Wilhelm Wien (1864–1928) produced a model that was correct only for radiation with short wavelengths. However, in 1900 another German physicist, Max Planck (1858–1947), found a model that described the entire spectrum. Planck assumed that light was emitted only in discrete bundles (quanta or photons), each with energy $E = h\nu$ (where ν is the frequency of the light and h is a new constant equal to 6.626×10^{-34} Js).

This Hubble Space Telescope image shows embryonic stars emerging from a nebula about 7,000 light-years from Earth. The telescope has been used to study celestial objects that are too distant for ground-based telescopes to observe.

A second problem confronting physicists at the beginning of the 20th century was the manner in which electrons are ejected from metals when the metal surface is illuminated by light (see PHOTOELECTRIC EFFECT). To explain the photoelectric effect, German-born U.S. physicist Albert Einstein (1879–1955) used Planck's quantum hypothesis. He assumed that light photons behave in a particle-like fashion, each carrying energy $h\nu$.

The wave theory encountered further difficulty when a series of experiments by U.S. physicist Albert Abraham Michelson (1852–1931) and others failed to detect Earth's motion through the ether, in which light waves supposedly vibrate. Experiments indicated that light always travels at the same speed regardless of the relative motion of the source and the receiver. Einstein conclusion that there is no ether and no preferred frame of reference led him to develop his theory of relativity in 1905 (see RELATIVITY).

Quantum electrodynamics

The modern theory of quantum electrodynamics describes the strange nature of photons and electrons and has been found to agree with experiments to remarkable accuracy. Light acts so strangely that throughout history thinkers have been perplexed by its behavior. It is no wonder. The modern theory of quantum electrodynamics (see QUANTUM THEORY) shows that light consists of photons. The theory also shows that a photon explores every path when it travels through space. One cannot tell which path a photon will take. It is only possible to calculate the probability that a photon will take a particular path.

Modern optics

The first successful operation of a laser, in July 1960 by U.S. physicist Theodore H. Maiman (b. 1927) opened a new era in modern optics (see LASERS AND MASERS). The ready availability of bright, collimated, and coherent light gave rise to an explosion in the number of practical devices. Photographically recorded interference patterns resulted in three-dimensional holographic images (see HOLOGRAPHY). Lasers combined with progress in optical fibers created a revolution in communications (see FIBER OPTICS). The data on Optical disk media, such as compact discs (CDs) and digital versatile discs (DVDs), is read by shining a laser onto their reflective surfaces. Bar code scanners used in stores are anoth common optical tool.

A. WESTERN

See also: COLOR; DIFFRACTION AND INTERFERENCE; LASERS AND MASERS; LIGHT; NEWTON, ISAAC; PHYSICS; REFLECTION AND REFRACTION; TELESCOPE.

Further reading:

Hecht, Eugene. 2002. *Optics*. Reading, Mass.: Addison-Wesley.
Menn, Naftaly. 2004. *Practical Optics*. Boston: Elsevier Academic Press.

SOME MEASUREMENTS OF THE SPEED OF LIGHT

Experimenter	Date	Result	Method
Aristotle	330 B.C.E.	Light does not travel but is a quality of the medium, which it acquires all at once	
Galileo	1600 C.E.	If not instantaneous, light is extraordinarily rapid	Lanterns and shutters
Roemer	1675	125,000 mps (200,000 km/s)	Moons of Jupiter
Fizeau	1849	200,000 mps (315,000 km/s)	Toothed wheel
Foucault	1862	185,000 mps (298,000 km/s)	Rotating mirror
Newcomb	1882	186,342 mps (299,860 km/s)	Rotating mirror
Michelson	1926	186,285 mps (299,796 km/s)	Rotating mirror
Florman	1954	186,284 mps (299,795.1 km/s)	Radio interferometry
Woods et al.	1976	186,282 mps (299,792.4588 km/s)	Lasers
International	1983	186,282 mps (299,792.458 km/s)	Defined by international agreement

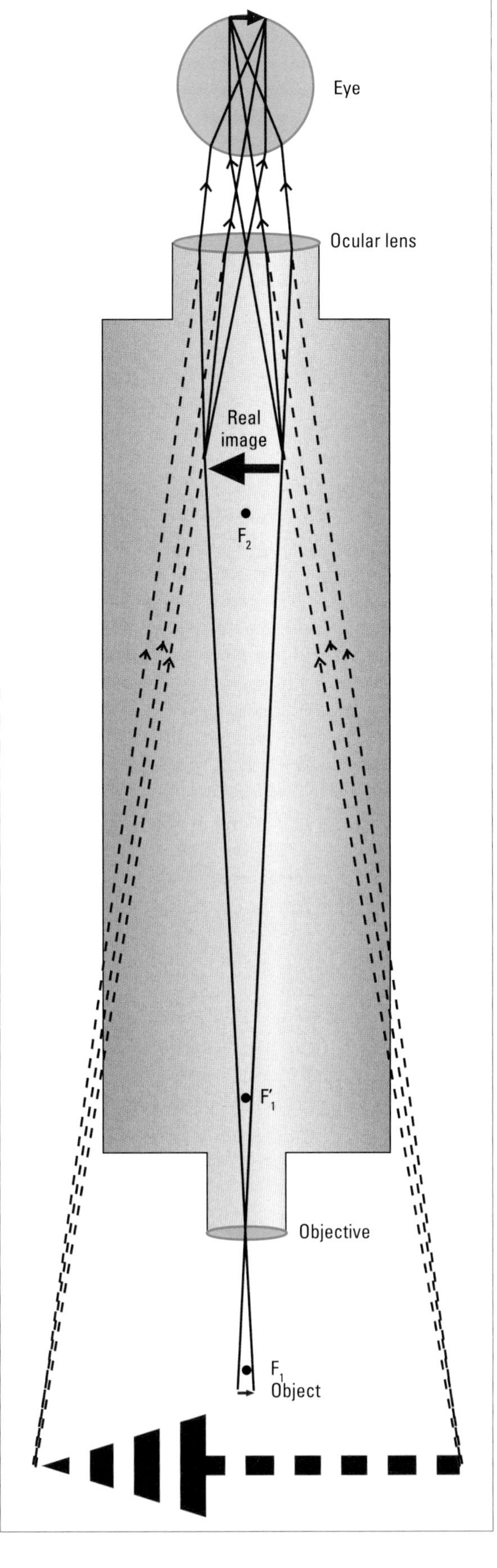

The compound microscope shown above has an objective lens and an eyepiece, or ocular lens, represented here by a single lens. A real image is produced by the objective lens, which then serves as an object for the ocular lens. A virtual image is produced by the ocular lens and is viewed by the eye.

ORDOVICIAN PERIOD

The Ordovician period is the geologic time period that lasted from 505 to 438 million years ago

The Ordovician period is the second oldest period of the Paleozoic era. It followed the Cambrian period, lasted 67 million years, and was followed by the Silurian period (see CAMBRIAN PERIOD; SILURIAN PERIOD). The system was named by English geologist Charles Lapworth (1842–1920) in 1879 for a Celtic tribe called the Ordovices, which lived in northern Wales before the Roman invasion of Britain. A commonly accepted practice was to name the geologic periods after ancient tribes that lived in the area where the rocks of each period were first described, as is the case for both the Cambrian and Silurian periods, as well as the Ordovician.

During this period, the continents looked nothing like they do today. For example, the entire eastern side of modern North America faced south, and the continent was situated in the tropics. North America was joined to Greenland to form a landmass called Laurentia. Other landmasses present at this time include Gondwana (comprising present-day South America, southern Europe, Africa, the Middle East, India, Australia, New Zealand, and Antarctica; see GONDWANA); Baltica (northern Europe and European Russia); Kazakhstania (central Asia); China (China and Malaysia); and Siberia.

Throughout the Paleozoic era, the various continents experienced constant flooding by shallow seas that encroached on the land, remained for millions of years, and then retreated. At the end of the Cambrian period, much of today's North American continent was covered by the fairly shallow Sauk Sea, which gradually withdrew during the first several million years of the Ordovician period, leaving exposed many limestone deposits (see LIMESTONE)—particularly throughout the American Midwest. During the middle of the period, about 460 million years ago, new flooding occurred by the Creek Sea, which led to extensive deposits of sandstone and limestone. The Creek Sea was the initial phase of a larger body of water, the Tippecanoe Sea, which covered much of the center of the continent. Deposits of limestone, sandstone, and shale from this period testify to the presence of this seaway.

CORE FACTS

- The Ordovician period was sandwiched between the Cambrian and the Silurian periods in the Paleozoic era and lasted from 505 to 438 million years ago.
- The landmasses present on Earth during the Ordovician period were Laurentia, Gondwana, Baltica, Kazakhstania, China, and Siberia.
- Ordovician rocks include limestones, sandstones, shales, and mudstones.
- Ordovician life encompassed corals, graptolites, crinoids, trilobites, brachiopods, and mollusks.

Ordovician rocks are found at Earth's highest elevations, such as the Ordovician carbonates underlying Mount Everest.

Ordovician rocks

Ordovician deposits can be found in nearly all parts of the world. One interesting aspect of such deposits is the fact that Ordovician rocks (known as the Ordovician system) have the distinction of forming Earth's highest elevations.

Geologists have noted that rocks formed from sediments deposited on the edges of Ordovician continental shelves (see CONTINENTAL SHELVES), as well as in the deeper ocean basins, are commonly dark, organically rich mudstones. In some basins, volcanic-generated igneous rocks are interbedded in the mudstones. These dark mudstones are often thinly layered and contain the fossilized remains of tiny organisms such as plankton and algae. Some rocks contain thin seams of iron sulfide.

Throughout the tens of millions of years of the Ordovician period, the world's landscape was nothing like it is today, and great shallow seas covered

CONNECTIONS

- The landmass **GONDWANA** was present during the Ordovician period.
- Flooding during the Ordovician period led to extensive deposits of **LIMESTONE**.

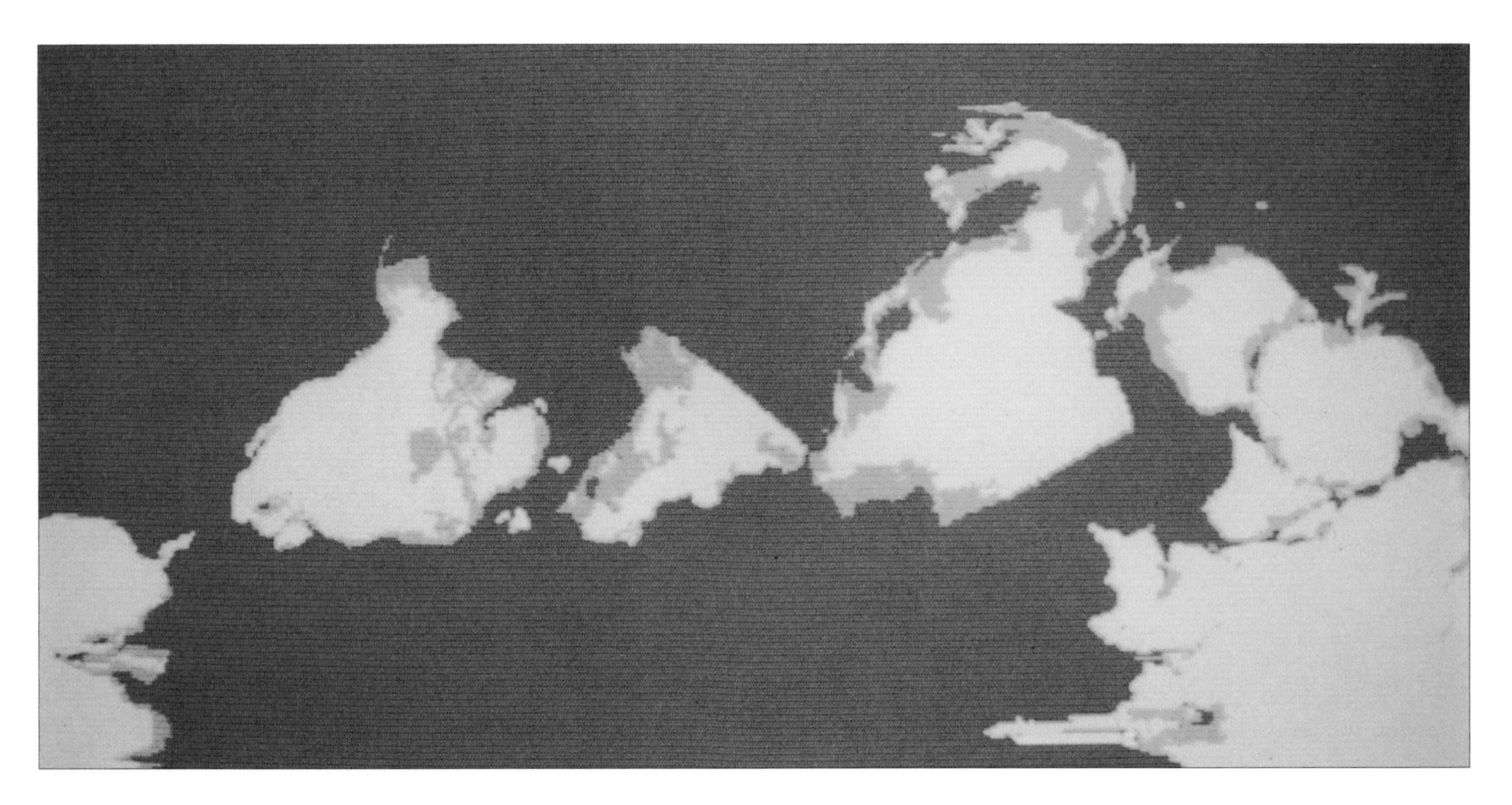

This computer illustration shows the position of the continents 500 million years ago.

most of North America. Most of Earth experienced a warm, humid—often tropical—climate. However, during the late Ordovician period, things began to change. On the basis of fossil evidence, it is believed that the equator may have stretched diagonally across North America from the Gulf of California to Hudson Bay, and in what is now the hottest part of the Earth, the vast Sahara, the temperatures dropped low enough for an ice age to begin and for glaciers to form (see GLACIERS AND GLACIOLOGY; ICE AGES). These processes are not so difficult to understand when one remembers that what is now central and northern Africa was not located along the present equatorial belt but over the South Pole.

The Ordovician period, like other geologic periods of the Phanerozoic era, is divided into a number of series, on the basis of types of fossils contained in the sedimentary layers of rock. The earliest deposits form what is known as the Ibex series and include sandstones and limestones. This period was followed during the Middle Ordovician by the Whiterock series, which is revealed on the borders of the Adirondack Mountains by a remarkably pure, wind-transported and water-laid quartz sand, which formed the St. Peter sandstone from central Michigan through Illinois and along the lower Mississippi River basin. Toward the end of the period, the Mohawk series was laid down. These deposits are most characteristically carbonates less than 1,000 feet (300 m) thick running northward into Canada from Virginia and eastern New York. In central Canada, these carbonates lie directly atop Precambrian rocks and cover nearly all the crystalline rocks of early Paleozoic origin.

During the final stages of the period, the Cincinnati series was deposited. The early Cincinnatian is sandstone and shale deposited in a U-shaped trough stretching from the interior of the continent through the present states of Ohio, Kentucky, and Indiana. Deposits during this stage were predominately fluvial (that is, carried by rivers) sands and gravels as well as red silts and clays, which are most common in the east.

Ordovician fossils

The warm, humid climate combined with the extensive seas spread over the land allowed for the development of plants and animals. It is easy to define the Ordovician system because of the presence of a large range of identifiable fossils. Although life was restricted to the sea, a wide variety of invertebrates developed during the period; identifying geologic strata and age by the fossil remains of such creatures has made the geologist's task much easier.

Ordovician life

Life during the span of the Ordovician period was very different from that of the previous Cambrian period. During the Ordovician, there was a far greater diversity of shelled marine invertebrates and bottom-dwelling suspension feeders, including the brachiopods. During this period, the first coral reefs appeared. Although many new groups of marine invertebrates such as brachiopods, trilobites, mollusks, and graptolites evolved, many other animals that developed during the previous Cambrian period became extinct, including certain species of trilobites. With their demise, even more groups were able to evolve. Of the large number of new species of trilobites, brachiopods, and sea lilies that developed, many became part of a second mass extinction that occurred around 440 million years ago, at the end of the Ordovician period. In addition to the greater diversity of life in the Ordovician, there was also a much higher level of ecological complexity and of increased zonation of life above and below the seafloor than in any previous time in geologic history.

CHARLES LAPWORTH

As a boy growing up in Oxfordshire, England, Charles Lapworth (1842–1920) loved roaming the meadows around his home, observing the ever-changing scene. He brought home birds' nests and colorful stones. He enjoyed reading about nature and often wrote poetry based upon his observations. He ended up becoming a scientist of the highest order.

Charles Lapworth was born in Farindon, Berkshire, on September 20, 1842, and spent his formative years in the English Midlands. His education was purely literary, but his growing interest in the natural world led to a genius for geological research. At the age of 22, he accepted a position as schoolmaster in southern Scotland. During his vacations, he continued his boyhood hobby of walking the moors and collecting rocks and fossils. In 1881 he was named to the newly established chair of geology at Mason College (now the University of Birmingham, England), where he combined his love both of academia and geological research.

Lapworth hiked the Southern Uplands of Scotland, studying and mapping the various outcroppings of bedrock. His collection of a group of fossils known as graptolites, extracted from black shale bands, were used to date and trace the early history of the terrain around him. He was able to date much of the area to the Lower Paleozoic era. In an area near Moffat, in southern Scotland, he was able to show that the many bands of shale outcrops were actually a series of overfolds of a continuous period and not a multitude of different series as had previously been thought.

With his own expanding collection of graptolites, Lapworth tabulated a series of 20 zones, which he defined and dated. This mapping led, in 1889, to the most comprehensive account of the stratigraphy of the Southern Uplands. The two published findings of his research and dating methods, *Geological Survey Memoir on Southern Scotland* (1899) and *Monograph of British Graptolites* (1901), inspired additional research into the use of graptolites to date and map the stratigraphy of other British regions.

During the late 19th century there was a major debate among geologists concerning the division of the Lower Paleozoic rocks: those located beneath the Devonian Old Red Sandstone beds. Scottish geologist Sir Roderick I. Murchison (1792–1871) held the belief that only the Silurian beds encompassed the Lower Paleozoic, while English geologist Adam Sedgwick (1785–1873) argued that these beds should be divided into those belonging to two periods: the Cambrian and the Silurian. In 1879 Lapworth, in an attempt to mediate the dispute, proposed a major classification of the Lower Paleozoic rocks into three systems, rather than trying to fit the sequences into one or the other. He noted that three distinct faunas had been recorded in Lower Paleozoic beds and observed that each was as "marked in their characteristic features as any of those typical of the accepted systems of later age." His suggestion of a new system, the Ordovician, was a perfect solution to the Sedgwick-Murchison dispute and was accepted almost immediately.

Lapworth continued his investigations of the strata of the Southern Uplands, providing major contributions of the age and character not just of Scotland but of the entire British Isles. He mapped much of the Cambrian beds in Wales and the Precambrian, Cambrian, and Ordovician beds throughout Britain. From his Birmingham base, he discovered several Lower Cambrian fossils, including the Olenellus trilobite, which became an important standard in identifying many of the Cambrian and Ordovician beds. His original work led to the Royal Medal by the Royal Society in 1891 and the Geological Society's highest honor, the Wollaston Medal, in 1899. Charles Lapworth died in 1920 at the age of 78.

DISCOVERERS

A rare fossil specimen of the starfish* Lapworthura miltoni*, which originated from the Ordovician period, approximately 450 million years ago.

This mixture of Ordovician fossils, found in Shropshire, England, dates back 435 to 500 million years ago and includes brachiopods and crinoids.

Pelecypods (clams), which had first appeared in the Cambrian period, became more diverse in the Ordovician. These mollusks were shortly followed by their deadly enemy, the starfish. Starfish devour clams by wrapping their tentacles around the clams, drawing open the shell, and sucking out the body.

One of the more important invertebrate animals to develop during the Ordovician period was the vast array of corals. Corals come in a variety of types, but all develop a hard external skeleton made of calcium carbonate. The horn corals live separately, but others form colonies beneath the sea and live on the various nutrients they are able to glean from the surrounding waters. They group together, and when they die, their hard skeletons remain and provide a foundation for the next generation of corals. In time, these millions of tiny remains grow into large, extended reefs along the seabed. The corals that originated during the Ordovician period are important fossils for relative age dating (see FOSSIL RECORD).

Living alongside the corals were the sea lilies, or crinoids (from the Greek word for "lily"). Although animals, these creatures looked like flowers, with an extended stalk and a cuplike bell at the end. From the bell extended several dozen feathery strands. As these tentacles wafted in the seawater, minute particles of smaller life-forms were drawn into the sea lily. The stalk was usually rooted to the seafloor, and the feathery "fingers" moved with the flowing sea.

In addition to corals and crinoids, various forms of moss animals called bryozoans grew along the Ordovician seabeds. Although, like corals, they left a calcareous shell when they died, they were quite distinct anatomically from the corals. They looked like fuzzy antlers growing from a spreading root system. Actually, bryozoans were formations of thousands of tiny invertebrate animals growing together. The bryozoans and sea lilies, as well as five-armed starfish, were the first organisms to develop during the early Ordovician period.

Around 460 million years ago, during the Middle Ordovician, sea scorpions, or eurypterids, evolved. These strange creatures grew as long as 9 feet (2.7 m) and had streamlined bodies with paddle-shaped rear appendages like those of lobsters. Attached to a eurypterid's thumblike head were long, clawed forelimbs. Their shape made them excellent swimmers. It is believed that they swam on their back and lived in shallow, semienclosed brackish lagoons, rather than in the deeper, open seas inhabited by the other invertebrates of the time.

Swimming along the bottom of the shallow seas were the nautiloids, a form of mollusk with a variety of different shell shapes. Some had straight shells and looked like ice cream cones with tentacled heads popping out the top. Others had slightly coiled shells. Shells with patterns of brightly colored stripes or zigzags were also evolving. Over the years, the shells on many of the nautiloids became coiled. The pearly nautilus, also called chambered nautilus, found in today's oceans is a direct descendent of the nautiloids that lived hundreds of millions of years ago. The Ordovician nautiloids, however, were often much larger than their modern-day counterpart. They fed on jellyfish or bottom-dwelling creatures such as shellfish, trilobites, and various types of algae.

The trilobites

The most widespread of the underwater creatures were the trilobites. They started their evolutionary chain during the early stages of the Cambrian period, more than 550 million years ago, and evolved through the Paleozoic era, becoming extinct during the early Permian period, some 245 million years ago. During the Ordovician period, however, many new types of trilobites came into existence. Trilobites had jointed legs and long feelers on their head. They varied in size from tiny ticklike creatures to those about the size of a playing card. They had a hard casing of flexible segments over their backs and the ability to curl up like an armadillo if they sensed danger. Trilobites were bottom feeders and often burrowed through muddy seabeds eating algae, small plants, or tiny sea animals.

What made some of these sea creatures so unusual was their complex eyes, which were much like those of today's insects, such as the housefly's. Each of the two eyes were large, rounded structures composed of many individual lenses organized in a regular pattern. Each of the lenses was pointed in a slightly different direction. Some trilobites had the ability to view more than 180 degrees. Some species had as many as 12,000 lenses in each eye.

Among the earliest known vertebrates was a jawless, armored fish called an ostracoderm. Their fossil remains have been found in nearshore marine strata of Middle and Late Ordovician rocks in Australia, Bolivia, and western North America. It is believed that these unusual early fish were bottom detritus feeders.

Although there was just the very beginnings of life on the land during the Ordovician period, sea life grew tremendously. Many individual species would die out over the millennium, but not a single phyla that developed from the Ordovician has become extinct, and they are still with us today.

B. LEERBURGER

See also: CAMBRIAN PERIOD; FOSSIL RECORD; GEOLOGIC TIMESCALE; GLACIERS AND GLACIOLOGY; ICE AGES; PALEOZOIC ERA; SILURIAN PERIOD.

Further reading:

Fortey, Richard A. 2002. *Fossils: The Key to the Past.* Washington, D.C.: Smithsonian Institution Press.
Prothero, Donald R., and Robert H. Dott Jr. 2004. *Evolution of the Earth.* Dubuque, Iowa: McGraw-Hill.

EARTHBOUND RESOURCES

It is hard to realize that so many of the things people use daily date from the time before four-footed animals walked on Earth. In the United States alone, for example, some of the oil and gas that is used to power cars and heat homes originated in Ordovician rocks ranging from Kansas through Oklahoma and Texas. Lead and zinc mines in Missouri are of Ordovician origin, as are similar mines in northwestern Illinois, Iowa, and Wisconsin. Many of the limestone quarries, so valuable in the construction industry, originated when millions of calcium-clad invertebrate creatures died in the Ordovician seas that covered parts of North America. The great marble quarries in Vermont, the source of the white marble used in the Lincoln Memorial (pictured below), as well as the vast slate beds in Pennsylvania, all date back some 500 million years.

Sedimentary beds stretching from the Appalachians northward to Newfoundland are now mined for their important iron ore (see ORES). Yet valuable Ordovician earthbound resources are not limited to the United States. Several years after gold was discovered in California in 1848 (see GOLD), a similar gold strike occurred in Australia's Bendigo district outside Melbourne. These rich gold veins all were created during the Ordovician period. In addition to gold, Australia's marble quarries are also of Ordovician origin.

SCIENCE AND SOCIETY

ORES

Ores are metal-containing minerals that can be extracted from veins in rock

One of the most important copper ores is chalcopyrite (copper iron sulfite), a common mineral that yields the by-products gold and silver.

CONNECTIONS

- Ores are removed by surface **MINING** wherever possible.
- **SMELTING**, in which an ore is melted with a flux that bonds with impurities, is used to recover **METALS** from most ores.

Ores are metallic minerals usually found in fractures, called veins, in solid rock. An accumulation of ores is called an ore body, and a region in which many veins of ore are found close together is called a lode. Ores are also found on beaches and in streams in deposits called placers, which are produced by the erosion of ore from veins and its subsequent deposition.

To be considered an economic ore deposit, the minerals must be of sufficient quantity and quality to be extracted profitably. Many factors influence the value of an ore body. For example, the relative percentage of useful ore in a well-defined mineral deposit determines the grade of the ore: highgrade ore is rich in metallic minerals. Commercial-grade ore, on the other hand, need not be high-grade if the market value of the metal is high enough. Precious metals, such as gold, silver, and platinum, are mined from ores that have a fraction of a percent metal content. Iron and aluminum ores need to be about 30 percent to make mining profitable. Samples from an ore body are assayed to determine the economic value of the ore (an assay is a chemical and physical analysis of the composition and purity of the sample).

Ores frequently contain more than one metallic mineral. The rock and mineral matter in the ore that has no value is called gangue. The chemical structure of an ore mineral can be a factor in determining commercial interest in an ore: native metals, that is, elemental (uncombined) metals, or alloys of metals are valuable ores. An alloy is a mixture, a solid solution, of two or more metals. Gold, silver, copper, and a natural alloy of gold and silver called electrum are found as native metals. Most ores are sulfides (compounds of metal and sulfur) or oxides (compounds of metal and oxygen). There are also common carbonates (compounds of metal with carbon and oxygen) and silicates (compounds of silicon and oxygen).

The location of ore deposits is an important economic factor. How deep the deposit is in the earth, whether the deposit is on the seabed or land surface, the shape and size of the ore body, the character of the terrain around the deposit, and the accessibility of transportation all influence the profitability of even a high-grade ore body. Economic geology is the branch of geology that deals with the location and the evaluation of ore deposits.

Origins of ores

Ninety-one elements are found at or near the surface of Earth. Most of them are metals: only 21 of them are nonmetals. In order of abundance, 10 elements account for 99 percent of the mass. Of these, the nonmetals oxygen and silicon make up three-quarters of the mass of Earth's crust. Aluminum is the most abundant metal at 8 percent; 5 percent of

CORE FACTS

- Ores are veins of economically valuable metallic minerals within rock.
- The value of the metal and the location and dimension of the vein determine whether a mineral deposit will be a profitable ore body.
- Ore deposits are associated with major geologic events involving magma and hydrothermal solutions and with the weathering and sedimentation of regions formed by these events.
- Mined ore is concentrated by separating it from waste rock in processes called ore dressing, after which the metal is liberated from the concentrate by smelting.
- Geologists search for ore deposits by studying evidence of ancient volcanism and mountain-building activities in rock formations.

the crust is iron; 2 percent is magnesium. Other important metals are less than 1 percent of the mass. Fortunately, metals are concentrated in ore deposits relatively close to Earth's surface. Ore concentrations are the result of dynamic geological processes such as magmatic segregation (see below), hydrothermal action, weathering, and sedimentation (see SEDIMENTARY ROCKS; WEATHERING).

Magmatic segregation

Large ore deposits are associated with periods of change in Earth's crust. Very old deposits may be related to the growth of continents, as described in the accretion theory, which suggests that stable continental masses grew larger with major movements of portions of Earth's crust.

Magma, molten rock that forms below Earth's crust in the mantle, rises and melts its way into cracks in crustal rocks (see EARTH, STRUCTURE OF; MAGMA). If magma reaches the surface, it is called lava. Intruding magma cools and sometimes interacts with the surrounding rock (known as country rock), altering the composition of both the magma and the rock. As magma cools, crystals are precipitated in a process called magmatic segregation, forming concentrations of complex ore minerals in the cracks.

Different minerals deposit at different temperatures. Chromite ($FeCr_2O_4$), which is an iron chromium oxide and the chief chromium ore, separates at very high temperatures, often in layered deposits. Some of the layered deposits that form are rich in both iron and chromium.

Hydrothermal solutions

When magma rises and more minerals are crystallized out, the now water-rich melt is better described as a hydrothermal solution. Near Earth's surface, cinnabar (HgS), an ore of mercury, is one example of an ore deposited from hydrothermal solutions (see MERCURY, METAL). Hydrothermal solutions may also be the result of brine solutions (seawater trapped deep down in sediments), with metals dissolved in them.

Many ore deposits are the result of metasomatic replacement, the process by which existing minerals are transformed totally or partially into new minerals by the replacement of their chemical constituents. In other words, minerals in solution replaced minerals in the surrounding rocks during the formation of the ore body.

Weathering and sedimentation

At the surface of Earth, weathering—the altering of rock minerals by the action of water, oxygen, and carbon dioxide—may occur. The aluminum ore bauxite is described as a secondary mineral, because it is produced from earlier formed minerals called primary minerals by the action of weathering processes. Bauxite is not one mineral but a mixture of hydrated aluminum oxides. Aluminum ore is concentrated by groundwater weathering action in tropical environments.

Sedimentation, the deposition of insoluble minerals from a fluid, also takes place at Earth's surface. One ore produced by sedimentation is the banded hematite (Fe_2O_3) iron ore formations in the Hamersley Range of Western Australia and the Lake Superior region of North America.

Running water can deposit heavy metals mechanically in a placer. The primary source of a placer deposit, the mother lode, is found upstream in the vein of a rock. Chemical weathering of the vein releases the metal, which then moves downslope. The metals may be concentrated on the slope of the land, but more commonly, they reach the stream. In order for minerals to concentrate in one place, they must be dense, resistant to chemical weathering, and not readily susceptible to cleaving. Native gold placer deposits were found in California, the Yukon, and Australia. The iron-sand magnetite (Fe_3O_4) deposit in New Zealand is an example of a beach placer.

THE ALLURE OF GOLD ORE

Discoveries of gold have changed the course of history (see GOLD). The Incan Empire in the Andes Mountains of South America was destroyed by Spanish conquerors greedy for native gold. The Aztecs of Mexico met the same fate. In the 18th century, the discovery of gold deposits sparked gold rushes to several remote parts of the world, and California was forever changed in 1849 following the discovery of a placer gold deposit along the American River. In 1851 the discovery of gold in New South Wales, Australia, created yet another gold rush. Gold rushes also occurred in the Yukon Territory of Canada and in Colorado.

HISTORY OF SCIENCE

In this photograph of a copper mine the colored areas indicate the presence of minerals such as malachite (green), azurite (blue), and cuprite (red), all of which are secondary minerals of copper.

Metal nodules containing high concentrations of cobalt, copper, iron, nickel, and manganese litter the deep water of the Pacific Ocean (see PACIFIC OCEAN). The metals are thought to derive from magmatic fluids flowing onto the seabed and developing around a nucleus, such as a fish bone.

From ore to metal

The trip from ore body to free metal can be as simple as panning for gold in a stream placer deposit—or it can be difficult, dangerous, time consuming, and costly. In order to reach the rich hematite iron ore trough of western Labrador, in the subarctic of the Canadian shield, a 360-mile (580 km), multimillion-dollar railroad had to be built.

Wherever possible, ore is removed by surface mining (see MINING AND PROSPECTING). If an ore deposit is too deep in the crust, an underground mine is constructed. Either type of mining activity requires access to the site for the heavy equipment that is needed to excavate, hoist, load, and haul enormous quantities of earth materials. First, the ore is crushed mechanically. Then equipment is used to separate the useful ore from gangue by a series of physical and sometimes chemical treatments to concentrate the valuable mineral. The treatments are collectively called ore dressing. Gangue is the material in the ore deposit that has no economic value; one example of gangue is rock.

Separation of the mineral from the gangue is based on characteristics of the mineral that differ from those of the gangue. For example, magnetic separation is used for the ferromagnetic ores pyrrhotite (which contain ferrous sulfide) and magnetite. Specific gravity differences are employed in some cases using jiggling, often in water.

Flotation is also used, particularly with sulfide ores such as lead and zinc. In this process, pulverized ore is mixed with water and a flotation agent, an oil that attracts the mineral. Air is forced into the mixture to form a froth of mineral-coated oily bubbles, which are skimmed off and dried to a concentrated cake. The waste materials left after concentration of the mineral, called tailings, are discarded.

A unique technique for liberating otherwise unprofitable ores of copper is a bioleaching process, which uses the naturally occurring bacteria *Thiobacillus ferrooxidans*. The bacteria acts in acid rainwater to oxidize the sulfur in the mineral and free the copper. Bioleaching produces toxic runoff, but it is useful for some low-grade copper ore. The technique is being used experimentally with cobalt and gold ore that is bound to iron and sulfur.

The mineral-rich concentrate then goes through more processes before ore is converted to metal. Chemically combined metals have to be reduced to become elemental metal. (Reduction is the chemical opposite of oxidation.) Different recovery methods are used for different metals. Smelting (the melting of ore with another material called a flux, which bonds with impurities; see SMELTING) is used for most ores. Electrolysis (the production of chemical changes by passing an electric current through a solution; see ELECTROLYSIS) is used in the final separation of metals such as aluminum and copper.

Searching for ores

Geologists look for mineral deposits by studying rock formations. Many mineral deposits are associated with ancient hydrothermal activity and magmatic segregation. Native gold is found with white quartz in veins of hydrothermal origin and downstream in placer deposits. Galena—lead sulfide (PbS)—the chief ore of lead, is associated with low- and medium-temperature hydrothermal deposits in veins where there are open cavities. Sphalerite—zinc sulfide (ZnS)—is frequently associated with galena, as is chalcopyrite ($CuFeS_2$, copper iron sulfide), which is an important copper ore.

Copper deposits in igneous rock are found along the Pacific coast of North and South America. Geologists theorize that these deposits are associated with the convergence of eastern Pacific plates with the American continents (see PLATE TECTONICS). There are similar mineral deposits on the opposite side of the Pacific. Here, where the eastern and western Pacific plates are diverging, the result is an upwelling of magma under the ocean.

Plate tectonics predicts types of mineral deposits in igneous rock that are characteristic of the volcanic and intrusive events that occur in mountain building, for example. Granites in the Himalayan continental collision zone have extensive tin-bearing mineral deposits of cassiterite, tin dioxide (SnO_2). Cassiterite is associated with high-temperature veins. Tin is mined in alluvial deposits (those formed from water-borne sediments) in China.

M. NAGEL

See also: CORROSION; EARTH, STRUCTURE OF; GOLD; IGNEOUS ROCKS; MAGMA; MERCURY, METAL; METAMORPHIC ROCKS; MINERALS AND MINERALOGY; PACIFIC OCEAN; PLATE TECTONICS; ROCKS; SEDIMENTARY ROCKS.

Further reading:

Jennings, Terry. 2003. *Metal*. North Mankato, Minn: Chrysalis Education.
Lynch, Martin. 2003. *Mining in World History*. London: Reaktion.
Pirajno, Franco. 2000. *Ore Deposits and Mantle Plumes*. Boston: Kluwer Academic.
Robb, L. J. 2004. *Introduction to Ore-Forming Processes*. Malden, Mass.: Blackwell Publishing.

URBAN ORE

Like natural ores, urban ore is an aggregate that contains metal—usually elemental metal—that can be extracted economically. Another term for urban ore is scrap metal, and the processes used to extract the metal are called recycling. Natural ores (sometimes called virgin ores) are nonrenewable resources; mineral ores are being mined faster than they are regenerated. Except for volcanic activity, geological processes that produce ores are slow.

The recovery of metals through solid-waste management saves energy and reduces the air and water pollution that result from the extraction of metals from virgin ores. The growing crisis of accumulating urban ore, increasing pollution, and the despoiling of land in heavily mined areas, as well as the need to use renewable resources, makes the use of urban ore more profitable.

Recycling urban ore can be fast. Aluminum cans can be recycled in six weeks using five percent of the energy that would be used if virgin aluminum ore had been processed. Other metals recycled profitably include iron, copper, and lead.

A machinery scrap yard in a California desert, an example of urban ore.

SCIENCE AND SOCIETY

ORGANIC CHEMISTRY

Organic chemistry is the study of carbon compounds

The organic chemist plays an important role in the synthesis of many medicines, fuels, plastics, and biologically active materials, all of which have had a profound effect on the quality of people's lives.

CONNECTIONS

- **FUNCTIONAL GROUPS** are particular arrangements of **ATOMS** at specific sites in molecules.

CORE FACTS

- Over 90 percent of all known chemical compounds are organic.
- The vast number and variety of organic compounds is due to carbon's ability to bond to itself to form long chains and rings of carbon atoms.
- Carbon bonds to a number of other elements, including hydrogen, oxygen, nitrogen, sulfur, and the halogens.
- Organic compounds are classified into families determined by the functional groups present in the compounds. The primary family of organic compounds is hydrocarbons, and all other families are considered their derivatives.
- Most organic compounds will burn in the presence of oxygen—a fact that is used in combustion analysis, which determines the percentage of carbon, hydrogen, and other elements present in organic compounds.
- Many modern techniques in organic analysis rely on computer technology.

Organic chemistry is associated with virtually every aspect of life. Many foods, medicines, cosmetics, and fabrics consist of organic compounds. Energy needs are met through the combustion of organic compounds contained in petroleum, coal, and wood. Almost all the chemical reactions necessary for life involve organic compounds.

During the 18th century scientists began to make a systematic study of all the chemical substances they could find. It seemed to them that there had to be a fundamental difference—some sort of life force—that distinguished compounds derived from living organisms from those obtained in the inanimate world. They made a clear distinction between organic—literally meaning "derived from living organisms"—and inorganic compounds. However, in 1828, one young German chemist, Friedrich Wöhler (1800–1882), found that heating ammonium cyanate—a salt prepared from inorganic sources—turned it into urea, an organic substance obtained from urine:

$$NH_4CNO \rightarrow NH_2CONH_2$$

This discovery opened chemists' eyes to the fact that organic substances could be analyzed, experimented with, and synthesized in exactly the same way as other chemicals. As a result, the field of organic chemistry expanded rapidly. Since almost all products of living things contain carbon, organic chemistry has come to be called the chemistry of carbon compounds. However, it should be recognized that although most carbon compounds are organic, a small number are considered inorganic, such as carbonates and carbides. Today millions of organic compounds have been identified, and tens of thousands are synthesized in the laboratory or isolated from plants and animals every year.

Carbon compounds

Carbon compounds are far more numerous than compounds of all the other elements combined. Well over 90 percent of all known chemical compounds are organic. Furthermore, most organic compounds have structures that are much more complex than most inorganic molecules.

Why is there such a vast number and diversity of organic compounds? The answer is found in the fact that each carbon atom has the ability to bind covalently to other carbon atoms. Therefore, carbon atoms form molecules of long chains or rings. They can also readily combine with atoms of a few other nonmetals, such as hydrogen, oxygen, nitrogen, sulfur, and the halogens (see NONMETALS).

Because of their structure, carbon atoms often form four covalent bonds (see CHEMICAL BONDS). Each atom has four electrons, which can be shared with an electron from another atom to make a pair. Thus, any carbon atom can bond to one, two, three, or four other carbon atoms. In this way a vast array of carbon structures is possible. For example, there are over 60,000 possible compounds containing only 18 carbon atoms and 38 hydrogen atoms.

In addition to forming single bonds, two carbons can form a multiple bond by sharing more than one pair of electrons. The simplest multiple bond is the double bond, which involves the sharing of two pairs of electrons by two atoms. In a triple bond, two carbon atoms share three pairs of electrons.

Organic compounds are classified into families that are defined by functional groups (see FUNCTIONAL GROUPS). These groups are particular arrangements of atoms, at specific sites in a molecule that are responsible for the physical and chemical properties of the molecule and determine most of its chemical characteristics. Members of a family can therefore be expected to react similarly. However, every part of an organic molecule can act as a functional group and have some influence on its properties, so this division into families recognizes that only one particular functional group can have a dominant influence.

Organic compounds can be simply defined as compounds with both a carbon-carbon bond and a carbon-hydrogen bond in the molecule. The simplest family of organic compounds is the hydrocarbons. As the name implies, hydrocarbons are organic compounds built of just carbon and hydrogen (see HYDROCARBONS). Molecules of hydrocarbons can be both chains and rings of carbon, with hydrogen at all the remaining bonding positions. Most of the organic compounds within families have names based on the hydrocarbon structures from which they are derived.

Organic analysis

Each new organic compound that is isolated from a natural source or synthesized in the laboratory must be analyzed to establish its chemical composition and molecular structure.

The first question the organic chemist asks is whether the compound is pure or a mixture of two or more other compounds. During the 19th century, when most compounds studied were relatively simple in their molecular structure, purity could usually be established by melting- or boiling-point behavior (see MELTING AND BOILING POINTS). A pure solid compound would have a definite melting point, and a pure liquid a constant boiling point. Impure mixtures generally have melting points or boiling points that are spread over a range of temperatures.

In addition, the unknown substance might form a compound with another organic or inorganic compound, such as an acid, which has a characteristic melting point or which forms crystals of a specific shape—another indication of purity.

The next stage is the determination of the composition (molecular formula) and the molecular weight of the compound. The standard method for determining the elemental composition of an organic compound is combustion analysis. This method involves the complete combustion (oxidation) of a carefully weighed sample of the compound in oxygen. The combustion converts all the carbon in the sample into carbon dioxide and all the hydrogen present into water. The combustion products are then absorbed— water by calcium chloride, and carbon dioxide by potassium hydroxide. The weights of the captured carbon dioxide and water are then used to calculate the percentages of carbon and hydrogen in the compound. If the figures obtained do not add up to 100 percent and no other elements are present in the compound, the difference is taken as the percentage of oxygen in the sample.

One may know the percentages of three elements present but not how many atoms of each make up the molecule. The molecular weight can be calculated by measuring how much a weighed quantity will raise the boiling point of a solvent such as benzene.

Developed by German chemist Justus von Liebig (1803–1873) in 1831, combustion analysis has been refined, particularly through miniaturization and

JUSTUS VON LIEBIG

During the formative years of organic chemistry, noted German chemist Justus von Liebig, shown below, laid the foundations for modern organic chemistry through his systematic study of organic compounds. His development of an analytical technique for determining the carbon and hydrogen content in organic compounds provided a tremendous boost to his work. Today's technique, commonly referred to as combustion analysis, is conceptually the same as Liebig's method.

In his later life, Liebig made important contributions in the areas of agricultural chemistry, where he advocated the use of mineral fertilizers, and biological chemistry, with his studies of nitrogen-containing organic compounds taken from plants and animals. A by-product of this work was a beef extract, which was sold as Liebig extract. Many of his students, including German chemists August von Hofmann (1818–1892) and Friedrich Kekule von Stradonitz (1829–1896), and French chemist Charles-Adolphe Wurtz (1817–1884), became notable chemists of the next generation.

DISCOVERERS

Advances in computer technology and the development of analytical techniques such as nuclear magnetic resonance and X-ray crystallography have dramatically changed the ways by which organic compounds can be identified.

automation, to the point where accurate and reliable analyses are possible using minute quantities of material—usually a few milligrams. In addition, today's method includes special techniques for determining the percentage composition of not only carbon, hydrogen, and oxygen but also other elements commonly found in organic compounds, such as nitrogen, sulfur, the halogens (fluorine, chlorine, bromine, and iodine), phosphorus, and certain metals.

The methods for determining the structure of an organic compound have changed dramatically in the last two decades with the development of sophisticated instrumental methods. Prior to this instrumental approach, the structure of an organic substance was determined, sometimes painstakingly, through information obtained from chemical reactions. The functional groups and the family to which a compound belongs could often be discovered by the use of special reagents. Some formed a characteristic color in solution; others converted the compound into an identifiable product or released simpler compounds, such as gases, that could be identified. These tests would be followed by chemically breaking down the compound into smaller fragments of known structure, from which a reasonable structure would be proposed. Proof of the structure was either secured by independent synthesis or by converting the original compound into a compound of known structure.

Instrumental methods have all but replaced the traditional approach for identifying the chemical composition of organic compounds. The most widely used instrumental methods for organic structure determination are infrared (IR) spectroscopy, ultraviolet (UV) spectroscopy, nuclear magnetic resonance (NMR) spectroscopy, and mass spectrometry.

Spectroscopic methods depend upon the way specific wavelengths of electromagnetic radiation are absorbed or emitted by atoms, functional groups, or molecules (see SPECTROSCOPY). In mass spectrometry, the sample is vaporized and bombarded with electrons, which break up the molecules into electrically charged fragments. These fragments are shot into a magnetic field, where their path is deflected into a curve. The amount of deflection is a measure of their molecular mass, and by this amount the different parts of the molecule can be identified.

Modern techniques

Instrumental methods of organic analysis have entered a new era with advances in computer technology. Elaborate computer programs allow instrument users to obtain spectra using milligram and sometimes microgram quantities of pure material. The classical methods of structure determination required (at least) gram quantities of material. Another advantage of the instrumental techniques is that they are almost all nondestructive—the sample essentially remains unchanged.

Most modern instruments are equipped with large databases of spectra that can be searched by computer. Databases contain from a few thousand to over 100,000 spectra. Once a spectrum is obtained for a compound on an instrument, it can be compared with spectra in the database in a matter of seconds. Spectra that most closely match the sample being studied are then reported. This information is invaluable in the determination of molecular structure.

Organic synthesis

The synthesis of organic compounds is an essential aspect of organic chemistry that has evolved tremendously since the synthesis of the first organic compound by Wöhler in 1828. The early approaches that were used to prepare relatively simple organic compounds have given way to sophisticated strategies for the preparation of organic compounds of considerable complexity. Organic synthesis has and continues to have a profound impact on people's quality of life. Many medicines, plastics, and biologically active materials are the direct result of the creative efforts of synthetic organic chemists.

The fundamental goal of organic synthesis is the preparation of a desired organic compound, either natural or designed, from readily available and relatively inexpensive starting materials. In practice, the synthesis of new organic compounds usually requires several—sometimes many—steps. In a multistep synthesis the product from one reaction serves as the reactant in the next reaction. Careful planning is required in a multistep synthesis so that it entails the fewest possible steps, because there is always a small percentage of loss at each stage. It is also desirable to use reactions that efficiently produce only a single product; avoiding the necessity of separating the desired product from a mixture of products, increases economy in both time and materials.

Much of organic synthesis involves the introduction of functional groups into a molecule and the interconversion of functional groups present in a molecule. Synthetic organic chemists have thousands of reactions at their disposal to accomplish these manipulations. Many organic reactions have been developed in just the last few decades. Major types include oxidation, reduction, substitution, addition, and elimination (see CHEMICAL REACTIONS). Most of the recent successes in organic synthesis can be attributed to increased understanding of these reactions through careful and methodical studies that explore the details of how they work.

Remarkable work continues to be performed in organic synthesis, with computers making a powerful impact (see COMPUTATIONAL CHEMISTRY). Sophisticated programs now exist that aid in the development of synthetic strategies for the construction of complex organic compounds. Another exciting area of research is the combination of biotechnological methods with traditional synthetic methods for the preparation of complex organic molecules.

Biotechnology makes use of both natural and modified enzymes to place a functional group in a specific place in a molecule or transform a functional group that already exists in a molecule into some other group. In some instances, it is possible to utilize several different enzymes to perform multistep synthesis in a single reaction vessel. Through genetic engineering, the number of synthetically useful enzymes is certain to increase, and thus, biotechnology will be a valuable addition to conventional methods of organic synthesis.

R. MEBANE

See also: ALCOHOLS AND ETHERS; ALDEHYDES AND KETONES; AMINES; CARBOCYCLIC COMPOUNDS; CARBON; CARBOXYLIC ACIDS; CHEMICAL BONDS; HETEROCYCLIC COMPOUNDS; HYDROCARBONS; ORGANOMETALLIC CHEMISTRY; PHENOLS.

Further reading:

Brown, T. L. et al. 2006. *Chemistry: The Central Science.* Upper Saddle River, N.J.: Pearson Education.
Carey, Francis A. 2006. *Organic Chemistry.* Dubuque, Iowa: McGraw-Hill.

COMBINATORIAL CHEMISTRY

One of the most innovative areas of organic synthesis is combinatorial chemistry. Primarily used by pharmaceutical companies in search of new drugs, combinatorial chemistry involves the rapid and simultaneous synthesis of minute quantities of a number of organic compounds with a similar basic structure. Current techniques use computerized automation systems to prepare hundreds to thousands of organic compounds per day. Traditionally, drug candidates had been synthesized one at a time, usually taking days or weeks to complete.

Called a library, the large number of related compounds generated by combinatorial synthesis are tested, often as complex mixtures, for bioactivity. Those members of the library showing promise are then isolated and identified for further drug development. The hope is that combinatorial chemistry will accelerate and revolutionize the process of drug discovery.

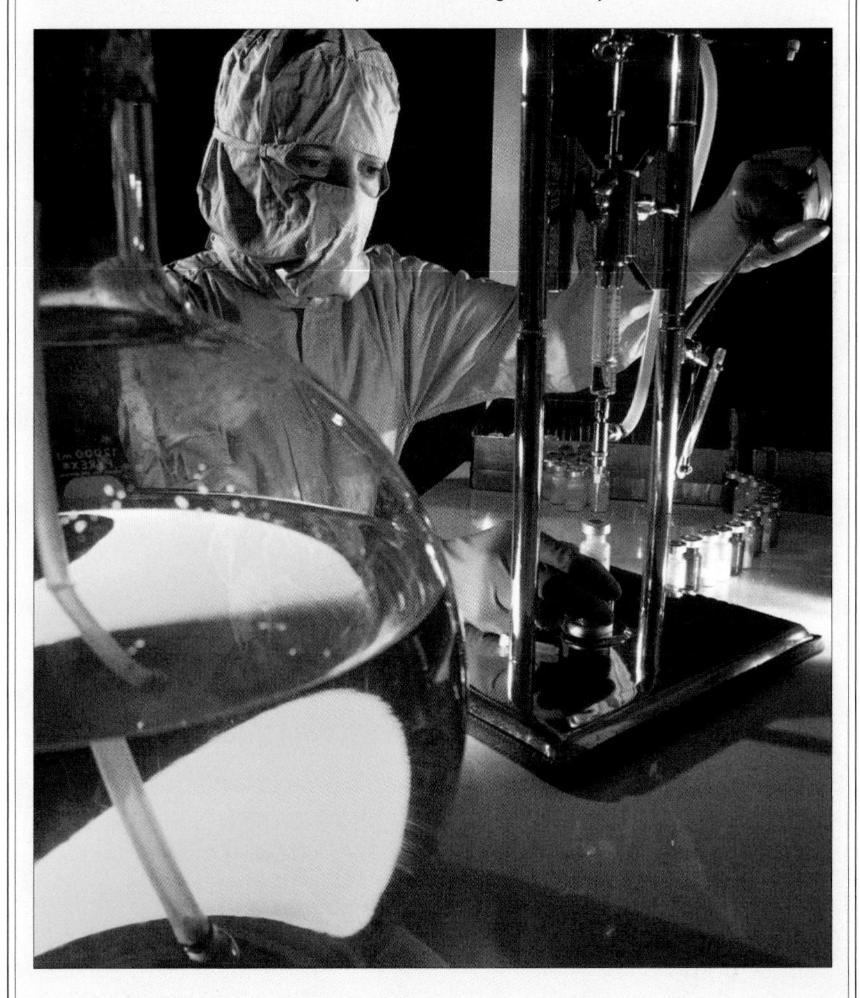

A technician prepares samples to test the sterility of conditions on a drug manufacturing line.

A CLOSER LOOK

ORGANOMETALLIC CHEMISTRY

Organometallic chemistry is the study of compounds that contain a metal bonded to carbon.

Organometallic chemistry is the study of compounds that all have one thing in common: at least one carbon-based ion, molecule, or molecular fragment (more commonly called a ligand) bonded through carbon to a metal atom (see CHEMICAL BONDS; ORGANIC CHEMISTRY). This definition gives rise to a huge range of different possibilities. Not only can the metal component be any of the s-, p-, d-, or f-block metallic elements, but a wide range of carbon-based ligands is possible (see METALS; PERIODIC TABLE).

Organometallic compounds

Organometallic complexes are classified according to the type of bond between the metal and the carbon atom. There are two main types of bonds. The first is a simple covalent metal-carbon (M–C) bond. They are just like the σ-bond between the two carbon atoms of ethane, C_2H_6, where each of the two H_3C units provides one electron in order to make up the carbon-carbon (C–C) bond (see the diagrams below).

The second class of bonds has the carbon ligand joined to the metal by a coordinate-covalent bond, just like the bond between boron trichloride (BCl_3) and ammonia (NH_3), in which ammonia donates its lone pair of electrons to the electron-deficient boron atom (see below; IONS AND RADICALS).

Common types of metal-carbon covelant bonds

M—C≡O metal carbonyl; M—C(R)(R)(R) metal alkyl; M—C₆H₅ metal phenyl; M—C(=O)R metal acyl; M=C(R)(R') metal alkylidene; M≡C—R metal alkylidyne

Common types of metal-carbon coordinate-covelant bonds

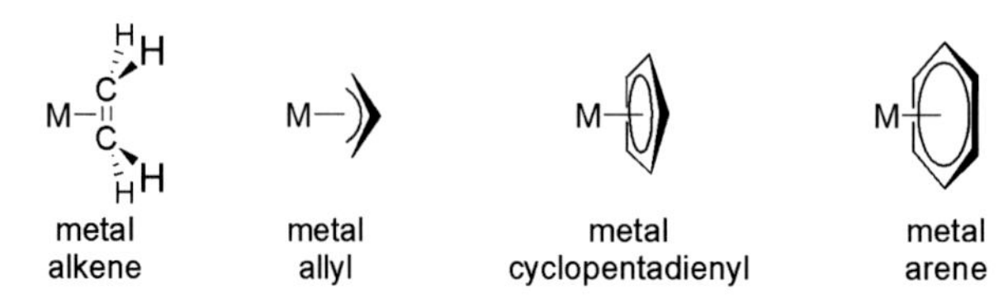

For organometallic compounds, this type of coordinate-covalent bonding occurs when unsaturated organic molecules or ions, such as those with a carbon-carbon double bond, for example, ethene (C_2H_4), donate a pair of electrons to an electron-deficient metal. In the case of ethene, it can donate the pair of π-bonding electrons from its double bond to a metal (see the diagram on page 1125).

Since benzene (C_6H_6) can be regarded as having three double bonds (in a so-called aromatic ring), it can donate up to six electrons. This organic molecule can therefore bond to metals easily, with the aromatic ring sitting like a crown over the metal (see opposite page).

Exactly the same idea can be used with ions, such as the cyclopentadienyl ion ($C_3H_3^-$), something that is described below. The fact that a whole range of different

CONNECTIONS

- Organometallic compounds show a wide range of different types of bonds between many different metals and **CARBON**-based fragments.
- Organometallic compounds are used as **CATALYSTS** for the synthesis of an enormous number of carbon-based products industrially.

Simple covelant bond

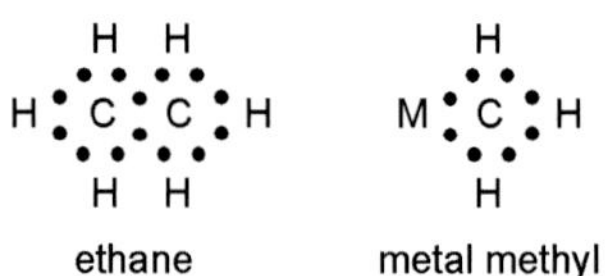

ethane

metal methyl

The two main types of bonds found in organometallic compounds. The M–C bond shown compares the similarity between ethane and a metal methyl.

Coordinate-covelant bond

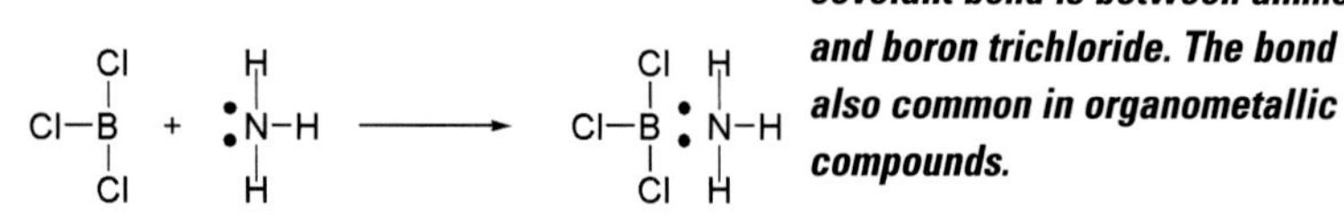

This illustration of a coordinate-covelant bond is between ammonia and boron trichloride. The bond is also common in organometallic compounds.

CORE FACTS

- Organometallic compounds possess a direct metal-carbon bond.
- The covalent metal-carbon bond of organometallic compounds is polar ($M^{\delta+}$–$C^{\delta-}$), a fact that explains the reactivity of these compounds.
- The bonding of carbon-based ions, molecules, or molecular fragments to a metal allows molecules with structures that are either unusual or even unknown in organic chemistry to be prepared and studied.

metal-carbon bonds can be made makes organometallic chemistry very diverse and exciting.

A defining characteristic of organometallic compounds is the polarization of their metal-carbon bonds (see POLARITY). The extent to which a particular metal-carbon bond is polarized depends simply on the differences in Pauling electronegativity between the metals and the carbon atoms (see ELECTRONEGATIVITY). The heavier s-block elements (sodium to francium; see ALKALI METALS), which are the most electropositive metals (or least electronegative), form M–C bonds that are ionic, because the difference in electronegativity between these metals and carbon is very great. The transition metals (p-block and d-block elements) are less electronegative (see TRANSITION ELEMENTS). These metals produced polarized M–C bonds, where the metal has a partial positive charge (written δ+) and the carbon atom a partial negative charge (δ–).

This M–C bond polarization is very important and helps to make organometallic compounds useful

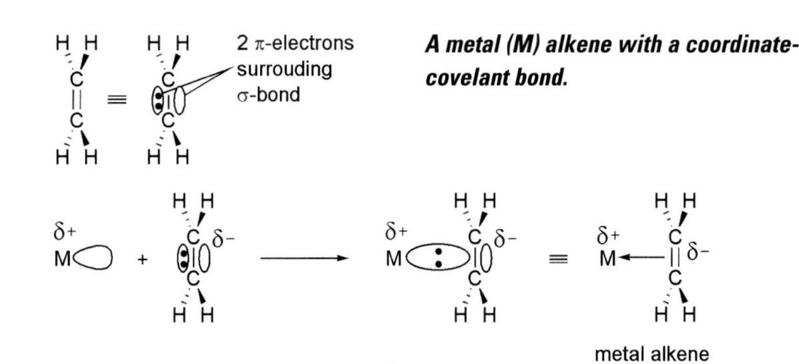

A metal (M) alkene with a coordinate-covelant bond.

Aromatic ring

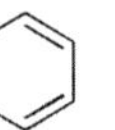

Metal atom and ring

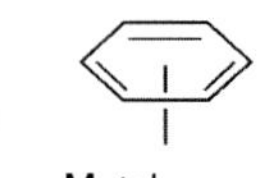

Metal arene

Benzene bonds to metal atoms to form metal arene. Six electrons are shared between the benzene ring and the metal atom.

EDWARD FRANKLAND

The British chemist Edward Frankland (1825–1899) was born in Catterall, Lancashire. He started his scientific career in 1840 as an apprentice pharmacist in Cheapside, Lancashire. In 1852, while working at Owens College Manchester (now the University of Manchester), he set out the fundamental concept of valency, which defines how a particular element may combine with a set number of atoms of a different element (the number of bonds which the central atom can form with other atoms). The so-called valency of a particular element is governed by the number of electrons in the outermost electronic shell of the atoms of that element—the valence electrons (see VALENCE).

In 1865 Frankland (pictured on the right), now professor of chemistry at the Royal College of Chemistry in London, was asked by the government to undertake an analysis of the quality of London's water and investigate means for its purification. This crucial work led to Edward Frankland being awarded a knighthood in 1897.

While studying the spectrum of light emitted by the Sun, independent work by French astronomer Pierre-Jules-Cèsar Janssen (1824–1907) and English chemist William Crookes (1832–1919) led to the first evidence of the existence of the element helium (see HELIUM). Later English chemist William Ramsay (1852–1916) isolated the first sample of helium in the laboratory from clèveite, a uranium-containing mineral. Confirmation that this gas was indeed helium was provided by the British astronomer Joseph Lockyer (1836–1920), who, with Frankland, gave the element its name after the Greek word *helios*, meaning "sun."

It is for his pioneering investigations of the reactions between metals such as mercury, zinc, arsenic, antimony, and tin with iodine-containing compounds of carbon (such as methyl and ethyl iodide) that Frankland is best known. This work led him to isolate more than 18 organometallic compounds and was carried out while studying for his doctorate (awarded in 1849) under the supervision of Robert Bunsen (1811–1899; known for inventing the Bunsen burner). Frankland's thesis was published in the Philosophical Transactions of the Royal Society in 1852.

Edward Frankland died a very rich man and is still highly regarded today for his outstanding scientific professionalism and his contributions to chemistry. He is often called the father of organometallic chemistry.

DISCOVERERS

reagents in other syntheses. The polar nature of the M–C bond opens up new types of reactions for the metal-bound carbon fragment that are simply not possible for a carbon-only compound. In some cases, it is only possible to have the organic component bound to a metal. Without the metal, they would not be stable or would be too reactive to exist for long.

Since carbon is known to be able to form single, double, and triple bonds (see CARBON), it is not unreasonable to suggest that these types of multiple bonds can also occur between a metal and a carbon fragment, as is indeed the case. A wide range of metal-carbon double (called metal-carbenes or alkylidenes) and triple bonds (metal-carbynes or alkylidynes) have now been prepared and their chemistry investigated in detail.

The origins of organometallic chemistry

A French scientist, Louis-Claude Cadet de Gassicourt (1731–1799), working in a Parisian military pharmacy, is believed to have made the first organometallic compound inadvertently in 1760. While experimenting on the preparation of invisible inks, he carried out a reaction that led him to isolate a bad-smelling liquid, which was highly toxic, smelled strongly of garlic, and spontaneously burst into flames when exposed to air. He called it his "fuming liquid." Since then, it has been shown that this product consisted mainly of cacodyl oxide ($[(H_3C)_2As]_2O$), named from the Greek words *kakodes* (evil-smelling) and *hyle* (matter). The feature that defines this compound as organometallic is the four single bonds between the metal arsenic and the carbons of the CH_3 groups.

$$As_2O_3 + 4\ H_3COOK \longrightarrow (H_3C)_2As{-}O{-}As(CH_3)_2$$

"Fuming liquid"

The next most important development was the discovery in 1825 by Danish chemist William Zeise (1789–1847) of the first transition metal organometallic complex of the metal platinum. After much painstaking laboratory work, Zeise was able to deduce the formula of his new compound (K, PtCl, $2H_2C$, 2Cl), although it was strongly contested at the time by the well-known scientist Justus von Liebig (1803–1873). Unfortunately, Zeise could not work out how these components joined together. German chemist Karl Birnbaum (1879–1950),

GRIGNARD REAGENTS

Organomagnesium halide reagents (RMgX, R=alkyl or aryl; X=Cl, Br, or I) are more commonly known as Grignards after their discoverer, the French chemist François Auguste Victor Grignard (1871–1935). The reagents are extremely important and versatile compounds, so much so that in 1912 Grignard was awarded a Nobel Prize for their discovery. He shared the prize with another Frenchman, Paul Sabatier (1854–1941), who had made advances in nickel-catalyzed hydrogenation of alkenes at about the same time.

$$>C{-}X + Mg \longrightarrow >C{-}Mg{-}X$$

Grignards play a vital role not only in organometallic chemistry, but also in organic chemistry. They readily react with water, oxygen, carbon dioxide, or almost any electrophilic organic compound (an electrophile is an "electron-loving" electron-deficient object that can accept a pair of electrons) and are routinely used in the synthesis of hydrocarbons, alcohols, carboxylic acids, and other such compounds. Their reactivity is easily understood, as they possess a polar covalent magnesium-carbon bond as a result of the difference in electronegativity between the two atoms: Mg = 1.3 and C = 2.5. As a result, the carbon component nucleophilic in character (a nucleophile is a "nucleus-loving" electron-rich object, which forms a bond by donating a pair of electrons).

An excellent example of the reactivity of Grignards is their use in the synthesis of alcohols. A Grignard will react with a ketone to produce a new magnesium salt following nucleophilic attack of the negatively charged carbon upon the ketone (see ALDEHYDES AND KETONES). The addition of aqueous acid to this magnesium salt gives the desired alcohol (see diagram below).

Although Grignards are most accurately described as having only a partial negative charge on the carbon fragment, it is nevertheless easy to imagine that such a carbon would be a very strong base, much stronger than needed to take a hydrogen ion (H^+) from water to generate the weaker base hydroxide ion (OH^+). Grignard reagents must be kept dry, away from even the slightest traces of moisture; otherwise they will be destroyed (often violently) by reaction with water. As a result, reactions with Grignard reagents are carried out under an atmosphere of an inert gas such as argon or nitrogen. Frankland originally used hydrogen, as this was the only dry gas he had available, in spite of any hazard owing to its flammable and explosive nature.

$$>\overset{\delta-}{C}{-}\overset{\delta+}{Mg}{-}X + R_2\overset{\delta+}{C}{=}\overset{\delta-}{O} \longrightarrow >C{-}CR_2{-}\bar{O}\ \overset{+}{Mg}{-}X \xrightarrow{H_3O^+} >C{-}CR_2{-}OH + \overset{+}{Mg}{-}X$$

Grignard — Ketone — Alcohol

A CLOSER LOOK

confirmed Zeise's, findings and went further, suggesting how a compound with this composition may form. However, it was not until 1951, at about the same time that the structure of ferrocene was being hotly debated (see box on page 1128), that the true identity of Zeise's salt started to become clear. English chemist M. J. S. Dewar (1918–1999) proposed that two CH_2 units and a metal could combine to give a compound in which a molecule of ethene ($H_2C{=}CH_2$) was bonded to the metal. This idea was followed up and applied to Zeise's compound in 1953 by English chemist Joseph Chatt (1914–1994), who suggested the structure below:

Zeise's salt

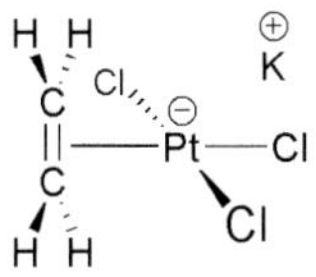

This idea was confirmed a year later, following an investigation by X-ray crystallography. This examination not only showed that Zeise had got the formula of his new compound entirely correct, but also that the two CH_2 fragments he proposed were indeed present as a molecule of ethene bonded directly to platinum. After 129 years the structure of this mysterious compound was finally solved, a discovery that helped establish organometallic chemistry.

The origins of organometallic chemistry as a discipline in its own right lie with the English chemist Edward Frankland (1825–1899; see the box on page 1125). While studying for his doctorate, he discovered that iodine-containing compounds of carbon (such as methyl and ethyl iodide [H_3C–I and H_3CH_2C–I]) react with various metals such as zinc, for example, if subject to heat or strong sunlight. The reactions gave new, low-boiling-point compounds that reacted violently or burst into flames on contact with moist air. In order to investigate the composition of these volatile and highly reactive new compounds, Frankland was forced to work under an atmosphere of dry hydrogen gas (a hazardous operation in its own right; see HYDROGEN), which prevented the compounds from coming in contact with air or water. Following painstaking work, Frankland deduced that he had made what he described as "organic bodies containing metals": dimethyl and diethyl zinc, $(H_3C)_2Zn$ and $(H_3CH_2C)2Zn$. They were the first organometallic compounds to have their true composition established beyond any further doubt.

Preparation of compounds

Generally, organometallic compounds are very straightforward to prepare. The simplest approach involves a direct reaction between a metal and an organo-halide (see HALOGENS). Two good examples of this type of synthesis are the preparation of alkyl lithiums, such as the synthetically and commercially import butyl lithium ($H_3CH_2CH_2CH_2C$–Li), and of Grignard reagents, R–Mg–X (see box on opposite page). These reactions are often exothermic, owing to the high enthalpy of formation of the salt (see CHEMICAL REACTIONS; THERMODYNAMICS).

Another common method of synthesis of organometallic compounds is through the replacement of the metal of an existing organometallic compound with a second different metal. There are a number of ways of preparing metal-carbon bonds using other organometallic reagents. One of the most versatile involves treating a metal halide with an organometallic compound.

Edward Frankland used all three of these synthetic approaches in his pioneering work on the identification and preparation of organometallic compounds carried out in the mid-19th century.

Organometallic compounds as catalysts

A catalyst is a substance that decreases the activation energy of a reaction so that the rate of reaction is increased. Transition metals have long been known to act as catalysts. A well-known example is the Haber-Bosch process, developed by Fritz Haber (1868–1934) and Carl Bosch (1874–1940), who were jointly awarded a Nobel Prize for their work in 1918. It takes nitrogen (an unreactive compound because of the strength of its N–N triple bond; see NITROGEN AND NITROGEN CYCLE) and combines it with hydrogen to form ammonia (NH_3). To bring about this reaction, it is necessary to pass the gases over an iron-nickel catalyst at high temperatures and pressures (752°F [400°C] and 200 atmospheres). There is no reaction at all when hydrogen and nitrogen are mixed without the catalyst (see CATALYSTS).

Today much of the success and interest in the study of transition-metal organometallic compounds can be linked to their use as catalysts for the manufacture of a number of useful chemicals. For example, 99-percent-pure acetic acid is prepared methanol and carbon monoxide using a catalyst system based on rhodium or iridium organometallic compounds and hydrogen iodide. The starting material, methanol, is also made using a copper catalyst.

Acetic acid is used for the preparation of vinyl acetate, can be converted into polyvinylacetate and employed in adhesives, in acrylic paints, and in the manufacture of latex. It is also converted into acetic anhydride, which is used as a solvent in the manufacture of cellulose acetate for films and fibers and to prepare aspirin (acetylsalycilic acid).

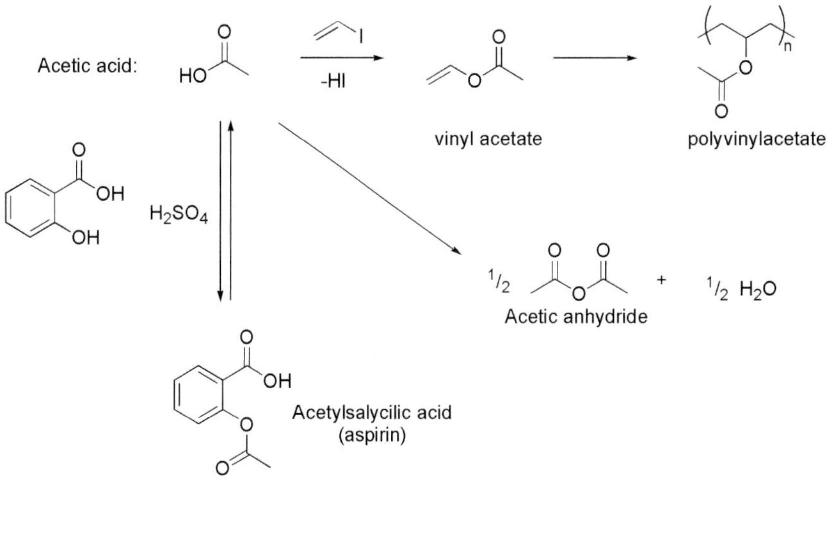

Acetic acid:
-HI
vinyl acetate
polyvinylacetate
H_2SO_4
1/2 Acetic anhydride + 1/2 H_2O
Acetylsalycilic acid (aspirin)

FERROCENE

The discovery of the compound, now commonly known as ferrocene transformed the way in which chemists think about the types of bonds possible between metals and ions and molecules based upon carbon. Ferrocene is an orange, crystalline, perfectly stable, iron-containing compound that does not react with water, has a melting point of 343°F (173°C), can be sublimed, has excellent solubility in most organic solvents, and has the simple empirical formula $FeC_{10}H_{10}$.

In 1951 two research groups, working independently of each other, isolated an unusual orange compound containing iron while they were trying to prepare the complex organic molecule fulvalene. Initially they thought that they had a simple compound in which two Fe–C covalent bonds linked the two hydrocarbon rings together (much as in Grignard reagents).

However, quite soon, the eminent German and British chemists Ernst Otto Fischer (b. 1918) and Geoffrey Wilkinson (1921–1996), working quite separately, realized that the highly unusual properties of ferrocene simply were not consistent with its being held together by two covalent bonds. Using chemical, physical, spectroscopic, and X-ray crystallographic analyses, they showed that only one structure could be used to account for all the observations that had been made. This was, quite simply, a double cone, or sandwich, structure, in which the iron atom was located above and below the center of the two hydrocarbon cyclopentadienyl rings (C_5H_5), which lie parallel to one another. For these groundbreaking, monumental pieces of deduction and experimental work, Fischer and Wilkinson were awarded the Nobel Prize in 1973.

3 (H, MgBr) + $FeCl_3$ → ✗ fulvalene; → [$(C_5H_5)_2Fe$] + $\frac{1}{2}$ $C_{10}H_{10}$ + 3 MgBrCl

The 'sandwich' structure of ferrocene: Fe = ○

What both Fischer and Wilkinson had realized was that the cyclopentadienyl rings (C_5H_5) were capable of bonding to iron (Fe) by forming both covalent and coordinate-covalent bonds at the same time. Put more simply, each of the carbon atoms present were all bonded to iron simultaneously; and thus, the 10 Fe–C bonds were all the same length.

The C_5H_5 rings viewed from the side and from above: Since all the C–C bonds were the same length in both of the rings, it was clear that in a particular ring, rather than having separate single and double C–C bonds (a carbon-carbon single bond is longer than a carbon-carbon double bond), the π-electrons and the negative charge form two rings of charge above and below each of the five-carbon skeletons. This situation is exactly the same as the one that is seen for the aromatic molecule, benzene (C_6H_6). These rings of charge could then interact with valence orbitals on the iron (like a crown being worn on a persons head). This symmetrical sandwich structure accounted perfectly for many of the properties of ferrocene.

For main group elements in the p-block, such as carbon and nitrogen, an octet of valence electrons is required to make them stable (s2p6). For transition metals (or d-block elements), a similar situation is seen, except that because these elements also have 5 d-orbitals, they require a total of 18 valence electrons (s2p6d10). This is called the "18-electron rule" and can be used to explain why ferrocene is so stable. Each of the cyclopentadienyl rings, which have 2 alkene bonds (2 electrons each) and a negative charge (1 electron), can give a total of 5 electrons to iron (which already has 8), making a total of 18, a stable valence state for a transition metal.

Together these observations opened the way for the preparation of a whole range of organometallic compounds in which the carbon ligands are bonded to metals using coordinate-covalent bonds as well as simple covalent bonds.

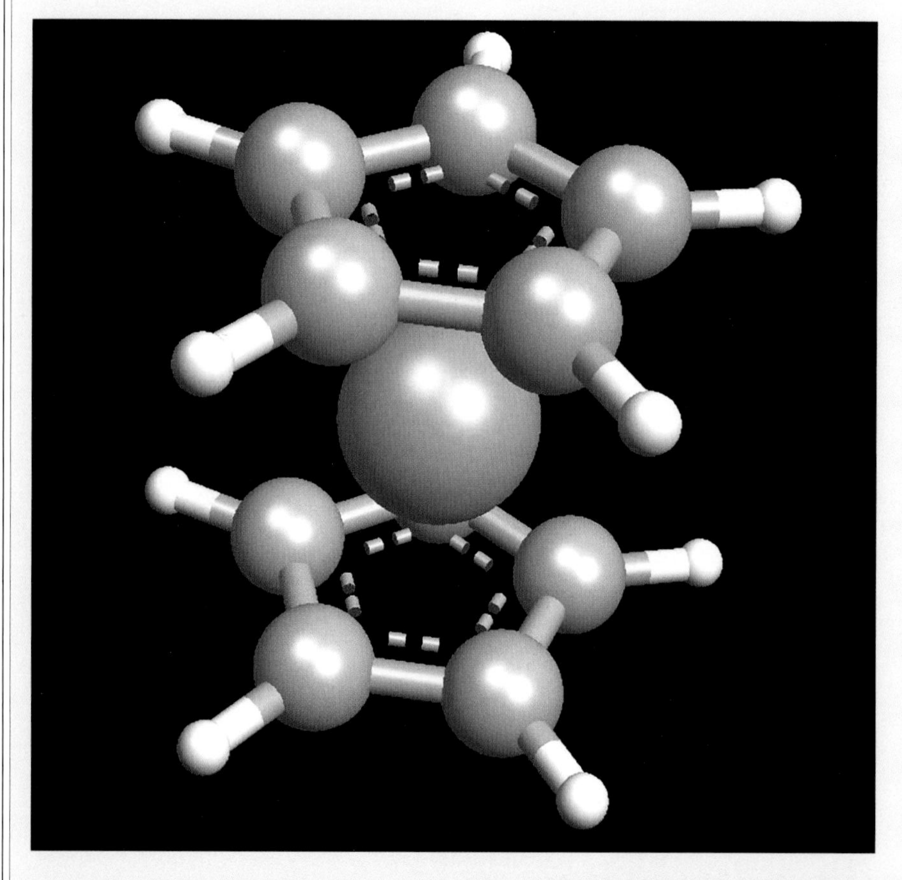

Ferrocene molecules have a complex sandwich structure with an iron atom between two rings of five carbon atoms.

A CLOSER LOOK

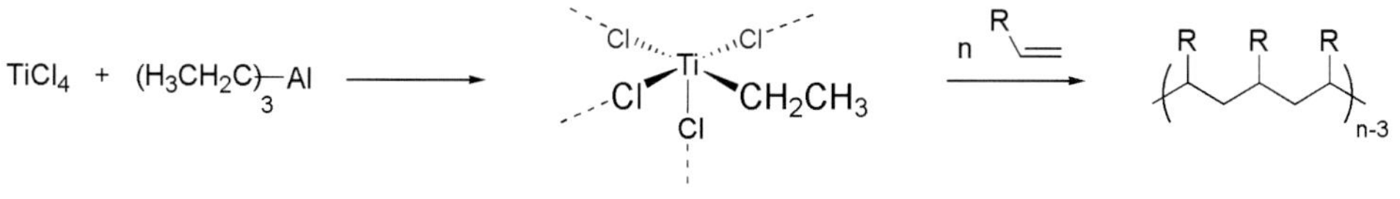

Ziegler-Natta Catalyst

One of the biggest areas in which organometallic chemistry has had an impact is the preparation of plastics, or polymers. Two of the most common plastics are polyethylene and polypropylene, which are made using organometallic catalysts. More than 55 million tons (50 million metric tons) of polyethylene are manufactured every year thanks to the groundbreaking work of Karl Ziegler (1898–1973) and Guilio Natta (1903–1979), whose contributions in this area led to their award of a joint Nobel Prize in 1963 (see PLASTICS). It was found that a titanium catalyst (now called a Ziegler-Natta catalyst) converted ethene and propene very efficiently, at low temperatures and pressures, to long-chain polymers that had excellent plastic properties (see diagram above).

Today organometallic catalysts are even used in the synthesis of complex pharmaceuticals, including the direct precursors to vitamin A and anti-inflammatory agents such as S-naproxen (related to ibuprofen) and cilastatin antibiotics:

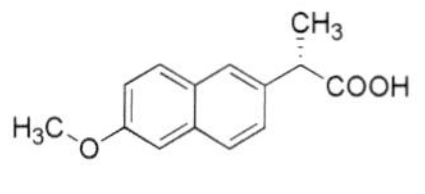

S-Naproxen

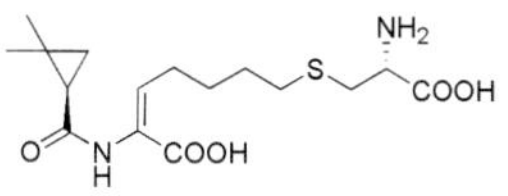

Cilastatin

Nature and medicine

It is important to realize that although organometallic chemistry is a comparatively new area of human study (ca. 150 years old), nature has long been making use of metal-carbon bonds in metalloenzymes. One of the best examples is Vitamin B_{12} (or cyanocobalamin), which is essential in the human body and has a Co–CN bond:

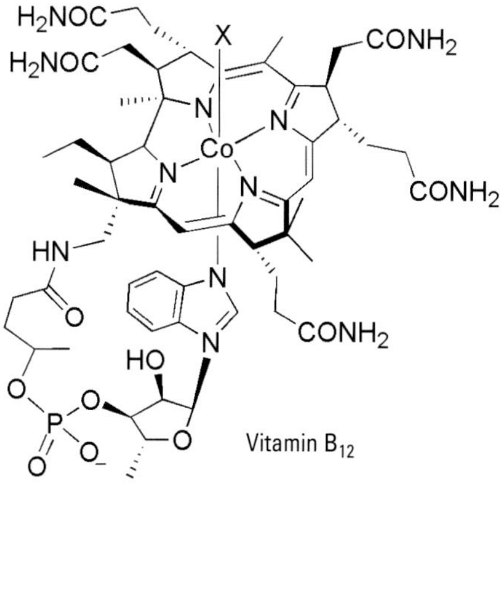

Vitamin B_{12}

B_{12} helps to maintain healthy nerve and red blood cells and is also needed to make DNA, the genetic material in all cells (see NUCLEIC ACIDS). People that suffer from the medical condition pernicious anemia, have an inability to absorb cobalamin naturally. This condition can lead to the potentially fatal formation of abnormal blood cells and is overcome by injections of large doses of cobalamin.

Nature makes good use of the complex atomic framework found in vitamin B_{12}. There are a number of related compounds that are also found in nature, such as the cobalt-based coenzymes. Coenzymes are substances upon which other enzymes depend for their activity. They are methylcobalamin, which possesses a cobalt-methyl bond, and cobalt-adenosyl, both catalyzing a number of different transformations of biological relevance.

One of the most fascinating uses of organometallic complexes is as antitumor agents:

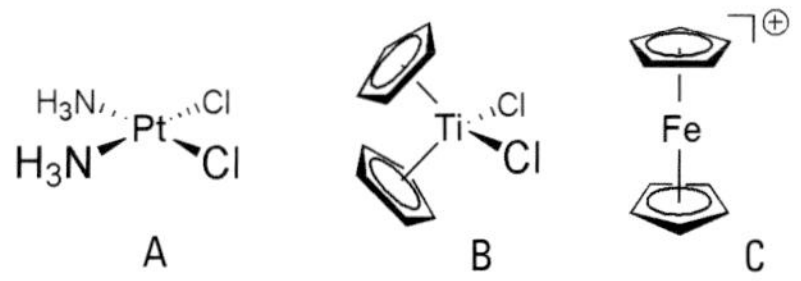

A B C

Cis-platin (A) is a metal-containing compound that has been used for over 30 years in the treatment of cancers. Although it is not an organometallic compound, since it does not have a metal-carbon bond, it is an excellent example of how effective metal compounds can be in medicine. Two organometallic derivatives, biscyclopentadienyl titanium dichloride (B) and ferrocenium salts (C) (obtained from ferrocene), have begun to be used to treat cancers.

P. DYER

See also: ALKALI METALS; CHEMICAL REACTIONS; ELECTRONEGATIVITY; HALOGENS; HYDROGEN; IONS AND RADICALS; METALS; MOLECULAR ORBITAL THEORY; NITROGEN AND NITROGEN CYCLE; NUCLEIC ACIDS; ORGANIC CHEMISTRY; PERIODIC TABLE; PLASTICS; POLARITY; THERMODYNAMICS; TRANSITION ELEMENTS; VALENCE.

Further Reading:

Carey, Francis A. 2006. *Organic Chemistry.* Dubuque, Iowa: McGraw-Hill.
Crabtree, R. H. 2000. *The Organometallic Chemistry of the Transition Metals.* New York: John Wiley.
Hill, A. F. 2002. *Organotransition Metal Chemistry*, Oxford, U.K.: Royal Society of Chemistry.

OSMOSIS

Osmosis is the net movement of water by diffusion through a selectively permeable membrane

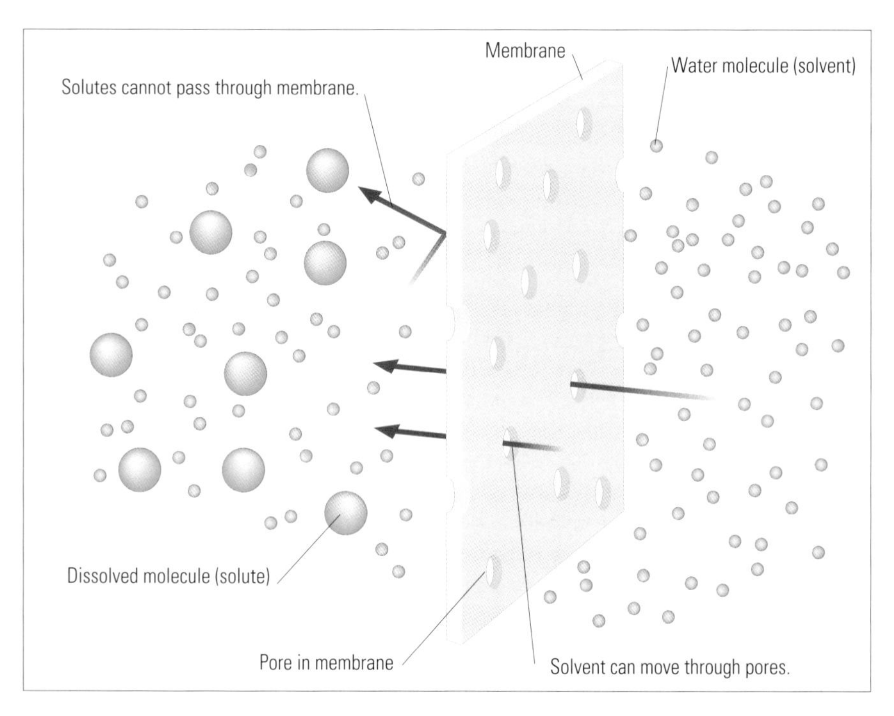

Osmosis is a form of diffusion that occurs when two solutions are divided by a membrane that does not allow large solutes to move through it. As this diagram shows, only the smaller solvent (water) molecules can move through the membrane. During osmosis, it is the water molecules that diffuse across the membrane to equalize the concentration of the substances dissolved in it.

CONNECTIONS

- Osmosis is a specific type of **DIFFUSION** that involves **WATER**.
- In osmosis, water diffuses through membranes from **SOLUTIONS** of low **SALT** concentration to solutions of higher salt concentration.

All living organisms are at least 50 percent water and require a regular supply of the liquid—even if only in their food—if they are to maintain their functioning bodies. The constituent cells of organisms—with the exception of some microbes—are bounded by membranes made of phospholipid and protein. These membranes are selectively permeable, meaning they allow some substances to pass through but not others. For example, small fatty (lipid) molecules tend to dissolve readily in membranes and so pass through relatively easily. Atoms or molecules that carry an electrical charge overall (called ions; see IONS AND RADICALS) or large molecules may be unable to cross a membrane. Water, on the other hand, is more or less freely permeable through selectively permeable membranes. The selective permeability of cell membranes is vital to enable an organism to maintain its body—and the cells within it—under chemical conditions that are distinct from its surroundings.

Osmosis is a specific type of diffusion. Diffusion is the random movement of small particles from one place to another until the particles are spread evenly throughout the region that encloses them (see BROWNIAN MOTION; DIFFUSION). Diffusion is the net movement of particles from regions of high concentration to those of lower concentration. Diffusion thus takes place down a concentration gradient. It is passive, meaning that it does not rely on externally applied energy. Diffusion continues until the particles are evenly distributed. The greater the difference in concentration between two areas (the steeper the concentration gradient), the faster the diffusion.

Osmosis is the diffusion of water through a selectively permeable membrane. Other factors aside, the rate at which water passes through a membrane is governed by the concentration gradient. If the gradient (the difference in concentration) is high, then diffusion is rapid; if the difference in concentration is small, then diffusion is comparatively slow.

A solution consists of solute dissolved in solvent (see SOLUTIONS AND SOLUBILITY). In a concentrated solution, there is a large amount of solute and a correspondingly low amount of solvent. A weak solution has a low concentration of solute and a correspondingly high concentration of solvent. In osmosis, water diffuses from a weak solution (with a high concentration of water) to a strong solution (with a low water concentration) across a membrane.

Osmosis and animal cells

Unless animal cells are surrounded by an isotonic solution (one that has the same concentration of dissolved solutes), water tends to move into or out of the cell by osmosis. If cells are placed in a hypotonic solution (one that has a weaker solute concentration than the cell contents), there will be a net movement of water into the cell by osmosis. Other factors aside, osmosis will continue until the solute concentration inside and outside the cell are the same, when osmosis halts. By that time, the cell will have swollen. If the original difference in concentration between the inside and outside the cell is great, the cell will swell so much that it bursts.

Animals and animal-like microbes called protists that live in freshwater environments have cell contents with much higher solute concentrations than their surroundings. They tend to take up water by osmosis. If their cells are not to burst, they must remove water as fast as it enters. Many animal-like protists, such as amoebas, do so using devices called contractile vacuoles, which are effectively water-baling mechanisms. The vacuoles fill with water and then discharge to the outside. This pumping mechanism consumes considerable amounts of energy but must continue as long as the water concentration gradient persists.

CORE FACTS

- Osmosis is important in the life of an organism because it is the means by which water usually enters or leaves cells.
- Other factors aside, water diffuses from a region of low solute concentration (and high water concentration) to a region of higher solute concentration (and lower water concentration).
- Plant cells and animals cells respond differently to osmotic inflows of water, because plant cells are bounded by a rigid cell wall whereas animal cells are not.

Osmosis and plant cells

The cells of bacteria, fungi, and plants have a rigid cell wall enclosing each cell. In plants, the cell wall is made of the carbohydrate cellulose, which acts as a rigid but highly porous scaffolding. It allows substances to pass through to reach the cell membrane, but it also prevents the cell from expanding too much. The normal state of a plant cell is slightly turgid, that is, with the cell contents pressing out against the cell wall.

If plant cells are placed in a moderately strong salt or sugar solution (hypertonic to the cell contents), water is drawn out of the cells by osmosis. The cells shrink away from their cell walls, a condition known as plasmolysis. The point at which the cytoplasm just begins to lose contact with the cell wall is called incipient plasmolysis. At that point, the cell contents no longer exert a pressure against the cell wall. In a living plant that does not have a tough woody stem, such loss of turgidity causes the plant to wilt (the stem and leaves droop). Wilting occurs on hot, dry, and breezy days when the rate at which the plant loses water through the leaves is greater than the rate at which the plant absorbs water from the soil through its roots. This imbalance causes cells in the leaves and stem to be temporarily starved of water, and they lose some of their turgidity, and hence the plant droops. In the evening, when temperatures cool, water absorption through the roots remains high while water loss through the stem and leaves declines. The plant stem and leaves become turgid again, and the plant stands upright.

As plant cells placed in a weak salt or sugar solution (hypotonic to the cell contents) take in water and expand, they press out against their cell walls with increased pressure. Eventually the pressure of the cell contents pressing against the cell wall (and the return pressure of the cell wall against the cell contents) is sufficient to resist the tendency for water to enter the plant cells by osmosis. At this point, the cells are fully turgid, and there is no net movement of water in or out of the cells. The plant cells do not burst, as is usually the case for animal cells placed in strongly hypotonic solutions.

OSMOREGULATION IN SALMON AND FRESHWATER EELS

Wild North American salmon and freshwater eels spend part of their life in freshwater and part of their life in the sea. Salmon eggs hatch in freshwater, the young migrate to the sea and grow to maturity there, and then, as adults return to freshwater to spawn. Freshwater eels hatch in the Sargasso Sea, near Bermuda (see OCEANS AND OCEANOGRAPHY), migrate into rivers and lakes to mature, and then return to the deep ocean trenches of the Sargasso Sea to spawn.

In both fish, the change from a freshwater environment to a salt water one involves a dramatic reversal of osmotic challenges. The concentration of a fish's body fluids lies between those of seawater and freshwater. In seawater, the body fluids are hypotonic relative to their surroundings. Water tends to leave the fish by osmosis through the gills and skin. To replace lost water, sea fish drink the seawater. They pump excess salts out across their gills, and their kidneys help conserve water by producing highly concentrated urine.

In freshwater, the body fluids of fish are hypertonic relative to their surroundings. Water tends to enter the fish by osmosis. Now the salt-pumping mechanism needs to be reversed, with fish pumping salts into their body across their gills. In freshwater, fish gain valuable salts from their food, and their kidneys produce large volumes of weak urine to get rid of excess water.

A CLOSER LOOK

Plants absorbing water from the soil

Most plants absorb the water they require from the ground. Typically, their roots have microscopic extensions called root hairs that increase the root's surface area for water absorption. Root cells normally have a higher concentration of solutes (and therefore a lower concentration of water) than the water in soil, so water diffuses from the soil and into the root cells. Moving across the root, from its outer edge to the inside, where transporting tubes called

This diagram shows how the net diffusion of water molecules into a hypertonic solution can exert enough force to push the solution up a thin tube. This force can be measured by using a weight attached to a piston. This same force is what drives water up a plant stem and allows water to reach the leaves atop even the tallest trees.

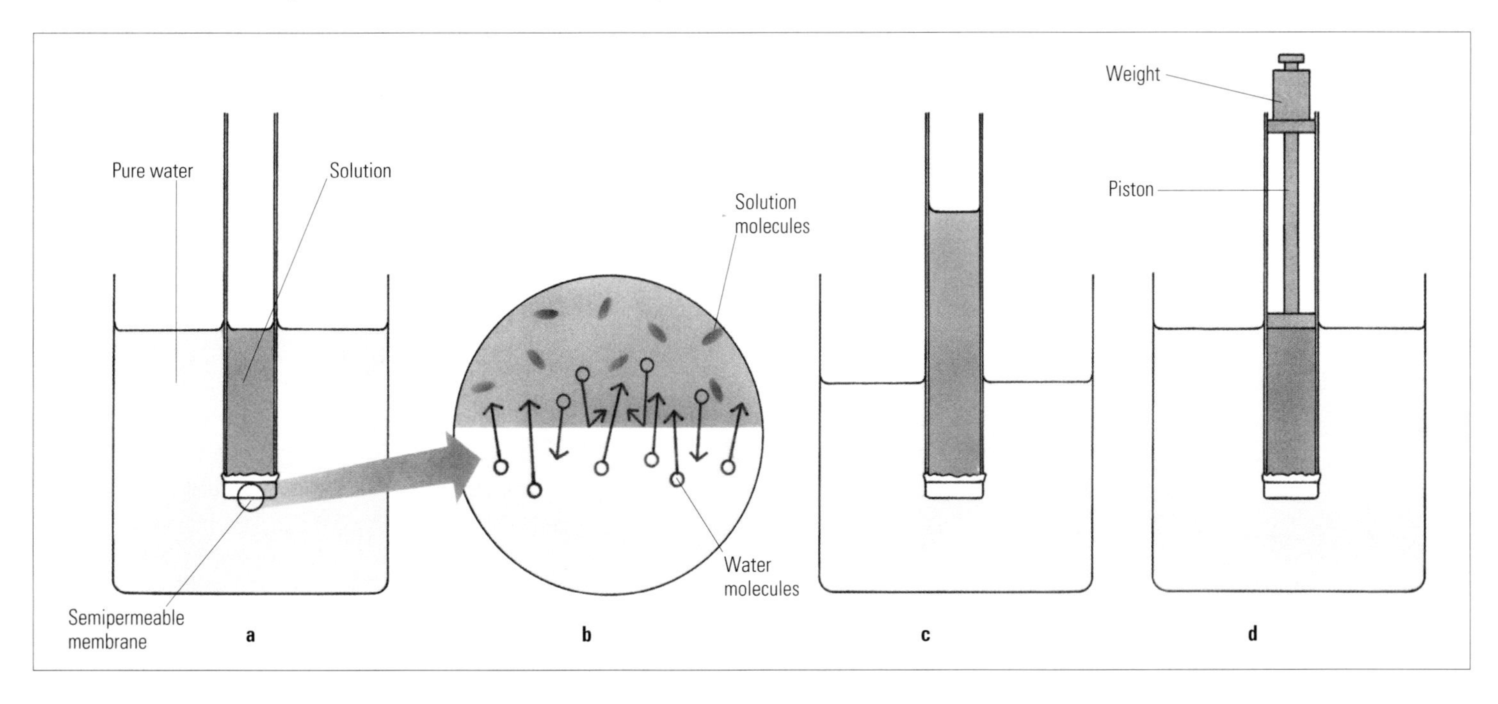

FOOD PRESERVATION

Many traditional methods of food preservation (see HALL, LLOYD), such as adding sugar to fruits to make jam and salting meat or fish, depend upon osmosis. Strong sugar or salt solutions tend to draw the water out of living organisms, particularly microbes and in doing so kill them or make them dormant. Placing foods in strong salt or sugar solutions helps preserve them from the actions of bacteria and fungi that would otherwise cause the foods to decay.

Fish is often preserved with salt. The salt dries the fish and makes it impossible for bacteria or other infectious microorganisms to survive.

SCIENCE AND SOCIETY

xylem vessels are found, there is a concentration gradient of dissolved solutes. The concentration gradient—a lower concentration of salts at the outside and a higher concentration to the inside—is set up and maintained by the energy-consuming active transport of solutes from one cell to the next. Water follows the movement of solute by diffusing down its concentration gradient.

There are three pathways that water can travel across the root: In the vacuolar route, water moves by osmosis from cell to cell and passes through the sap vacuoles of cells on the way. This route is probably the slowest. In the symplast route, water passes from cell to cell by osmosis and, within the cell, travels exclusively through the cytoplasm. In the apoplast route, water bypasses cells by traveling through the network of porous cell walls. However, even this water is forced to pass through cells when it reaches a layer of cells called the endodermis. In this layer, a waxy barrier called the Casparian strip blocks water passing through the cell wall and directs the water though the cell membrane and into cells. The apoplast route is probably the fastest route for water.

Whichever route water takes, it moves by osmosis at some stage. The Casparian strip ensures that all water passes through cells of the endodermis. These cells control the supply of dissolved solutes that travel from the roots to the rest of the plant.

Water travels from the roots up through the stem and into the leaves through a system of fluid-filled tubes called xylem vessels. In the leaves, water movement across the leaf tissue is largely by osmosis. For example, as water evaporates from the surface of leaf cells and escapes through pores in the leaf surface, water is drawn from the xylem vessels and into the leaf cells by osmosis.

The mammalian kidney

In vertebrates (animals with backbones), the kidneys are the main organs that control the balance of salt and water in the body's fluids (a process called osmotic regulation, or osmoregulation). Kidneys perform this function along with excretion—the expulsion of waste substances from the body (in the case of mammals, in the form of liquid urine). Kidneys work by initially filtering the blood so that water and small solute molecules enter microscopic structures called kidney tubules. Solutes are pumped out of the tubules and back into the blood according to the body's needs, and water follows the solutes by osmosis. The filtered liquid that remains in the tubules gathers in collecting ducts, where it forms urine. The urine passes out of the kidneys through ureters and is stored in the bladder before being expelled from the body through the urethra.

Among vertebrates, mammals have highly efficient kidneys. They can produce urine that is hypertonic (with a higher solute concentration than body fluids). In other words, mammalian kidneys are able to conserve water. They are able to do so because a solute concentration gradient exists between the outer part of the kidney (the cortex) and the inner part (the medulla). Within the kidney tubule, a section called the loop of Henlé concentrates salts in the medulla. When collecting ducts pass through the medulla, the high concentration of salts in the surrounding tissues draws water out of the urine by osmosis, the result being more concentrated urine.

Like other desert animals, the desert rat, or gerbil, rarely drinks but is able to obtain all the water it needs from its food or from chemical reactions within its body. The loops of Henlé in gerbil kidneys are particularly long, and the medulla of the kidney has a particularly high salt concentration. Gerbils conserve water by producing small volumes of highly concentrated urine.

T. DAY

See also: BROWNIAN MOTION; HALL, LLOYD; IONS AND RADICALS; OCEANS AND OCEANOGRAPHY.

Further Reading:

Burggren, W. W., K. French, R. Eckert, and D. J. Randall. 2002. *Animal Physiology: Mechanisms and Adaptations*. New York: W. H. Freeman & Co.

Campbell, N. A., J. B. Reece, and E. J. Simon. 2003. *Essential Biology with Physiology*. New York: Benjamin Cummings.

Hopkins, W. G. and N. P. A. Hüner. 2002. *Introduction to Plant Physiology*. New York: Wiley.

OXIDATION-REDUCTION REACTIONS

Oxidation-reduction reactions involve electron transfer from one ion, atom, or molecule to another

Oxidation-reduction reactions are electrochemical reactions in which both reduction (gain of an electron) and oxidation (loss of an electron) occur simultaneously. The electrons lost in an oxidation part of the reaction are gained in the reduction part. These reactions are also called electron-transfer reactions or redox reactions. They include many of the metabolic processes that take place within living organisms as well as the chemical reaction that occurs within a battery (see BATTERIES; BIOCHEMISTRY).

Oxidation states and oxidation numbers

Individual atoms are electrically neutral, being composed of negatively charged electrons and positively charged nuclei. The positive and negative charges within each atom balance each other out. When an atom loses or gains electrons, an ion is formed. The charge—or oxidation state—of the ion is the number of positive charges in the nucleus minus the number of electrons. Molecules and radicals can also exist as ions.

Chemists assign a formal charge, or oxidation number, to each atom in a molecule or polyatomic ion by applying the following rules:

1. The oxidation number of a free element is zero. (A free element is one that is not combined chemically with a different element, so O_2, H_2, and Cl_2, for example, are free elements and have an oxidation number of zero.)

2. The oxidation number of a monatomic ion is the same as its electric charge.

3. The sum of all the oxidation numbers of all the elements in a compound must equal the charge on the particle.

4. Oxygen has an oxidation number of -2 in its compounds, except in peroxides, where it has an oxidation number of -1.

5. Hydrogen has an oxidation number of $+1$ in its compounds, except in metal hydides, where it has an oxidation number of -1.

Rust (hydrated Fe_2O_3) forms on metal objects, such as this trawler, in an oxidation-reduction reaction. The metallic iron is oxidized to Fe^{3+} ions, while oxygen is reduced to O^{2-} ions.

As an example, these rules can be applied to a molecule of sulfuric acid (H_2SO_4), which is uncharged. Each hydrogen atom has an oxidation number of $+1$, and each atom of oxygen has an oxidation number of -2. The sulfur atom must have an oxidation number of $+6$ for the sum of oxidation numbers to be zero.

Oxidation numbers can be used to give another definition of an oxidation-reduction reaction. An oxidation-reduction reaction occurs when there is a change in the oxidation state of the atoms involved. The oxidation number increases during oxidation and decreases during reduction.

Oxidizing and reducing agents

The chemicals that participate in redox reactions are called reducing agents and oxidizing agents. A reducing agent loses electrons to an oxidized compound and is oxidized itself. The oxidized compound gains electrons and is reduced. Some substances, such as hydrogen peroxide (H_2O_2), can act as either a reducing agent or an oxidizing agent.

CORE FACTS

- Oxidation-reduction reactions play a big role in metabolic processes.
- The oxidation number of an atom is the formal charge assigned to the atom in the molecule or ion.
- Chemists assign oxidation numbers by applying a set of rules.
- Some substances, such as hydrogen peroxide, can act as reducing agents (donating electrons) or oxidizing agents (accepting electrons).
- The simplest type of redox reaction is one that occurs between metals and nonmetals.
- Photosynthesis is a process in which organic compounds are synthesized by the reduction of carbon dioxide in the presence of water and sunlight.

CONNECTIONS

- The transfer of **ELECTRONS** between the **ATOMS** of **METALS** and **NONMETALS** can result in a dramatic reaction.

PHOTOSYNTHESIS

One naturally occurring oxidation-reduction reaction occurs in all green plants. Photosynthesis is a process by which green plants, algae, and many types of bacteria reduce carbon dioxide to make nutrients: glucose (a sugar) and oxygen. In this process, part of the energy from the Sun is captured by a pigment called chlorophyll and stored as potential energy in glucose molecules. The reaction for photosynthesis is as follows:

$$6CO_2 + 6H_2O \rightarrow C_6H_{12}O_6 + 6O_2$$

The glucose produced from photosynthesis is the primary nutrient for all higher organisms. It provides the energy for driving cellular activity, as well as serving as a source of carbon for the synthesis of other important biochemical compounds such as lipids, amino acids, and nucleic acids.

Oxidation and reduction reactions are always paired: when something is oxidized, something else must also be reduced. In the equation for photosynthesis, the carbon is reduced from +4 to 0; the oxygen is oxidized from –2 to 0.

The energy stored in glucose is extracted and stored using many separate redox reactions linked in sequence. ATP (adenosine triphosphate) is a key compound involved in the transfer and storage of the Sun's energy. ATP is formed when an ADP (adenosine diphosphate) compound is linked to an inorganic phosphate. The bond between ADP and the additional phosphate group is capable of storing up to 8 kilocalories of energy. Once ATP is formed, it can be hydrolyzed in the reverse reaction to produce ADP and the phosphate, releasing the energy as needed.

Green plants make nutrients by photosynthesis, an example of a redox reaction.

A CLOSER LOOK

Balancing redox reactions

Redox reactions can broken down into two half-reactions to show which reactant is oxidized and which reactant is reduced. When sodium reacts with chlorine to give sodium chloride (table salt), the reaction is as follows:

$$2Na + Cl_2 \rightarrow 2NaCl$$

This reaction can be split two half-reactions:

Oxidation half-reaction: $2Na \rightarrow 2Na^+ + 2e^-$
Reduction half-reaction: $Cl_2 + 2e^- \rightarrow 2Cl^-$

In this reaction, sodium metal has lost electrons (has been oxidized) to form sodium ions. At the same time, chlorine atoms have gained electrons (have been reduced) to form chloride ions. The reaction illustrates a general point that applies to all redox reactions: the number of electrons lost in the oxidation half-reaction must equal the number of electrons gained in the reduction half-reaction.

Types of redox reactions

The simplest type of redox reaction occurs between metals and nonmetals. The transfer of electrons between the atoms of these elements dramatically changes their state. The reaction between sodium and chlorine to produce sodium chloride is an example of this type of reaction.

Another simple type of redox reaction is a combination reaction—two elements combine to form a chemical compound. One element is oxidized, and the other is reduced. For example, in the formation of water (H_2O), from oxygen and hydrogen, hydrogen is oxidized and oxygen is reduced:

$$2H_2 + O_2 \rightarrow 2H_2O$$

Oxidation: $2H_2 \rightarrow 4H^+ + 4e^-$
Reduction: $O_2 + 4e^- \rightarrow 2O_2^{2-}$

Another type of redox reaction is one in which an element replaces or displaces another from a compound. For example, when zinc metal is added to a solution of copper sulfate, the zinc replaces the copper ions to form zinc sulfate and copper metal:

$$Zn + CuSO_4 \rightarrow Cu + ZnSO_4$$

The sulfate ions are neither reduced nor oxidized and are termed *spectator ions*. The half-reactions are as follows:

Oxidation: $Zn \rightarrow Zn^{2+} + 2e^-$
Reduction: $Cu^{2+} + 2e^- \rightarrow Cu$

The zinc metal has reduced the copper ion to metallic copper, while the copper ion has oxidized metallic zinc to zinc ion.

M. RIESKE

See also: BATTERIES; BIOCHEMISTRY; CHEMICAL BONDS; CHEMICAL REACTIONS; CORROSION; ELECTROCHEMISTRY; ELECTROLYSIS; IONS AND RADICALS; OXYGEN.

Further reading:

Bartlett, Neil. 2001. *The Oxidation of Oxygen and Related Chemistry.* River Edge, N.J.: World Scientific.

OXYGEN

Oxygen, the most abundant element on Earth, is a highly reactive gas that is vital for respiration

Oxygen, in compounds with other elements, is the most abundant element in Earth's crust, of which it makes up some 50 percent by mass. In the oceans, oxygen accounts for 89 percent of the weight of water. In its elemental form, O_2, oxygen makes up nearly 21 percent of the atmosphere. Dissolved in water, it is vital to the respiratory processes of almost all living things, and it is also the most abundant element in living tissue. As ozone, O_3, it forms a layer in the upper atmosphere that absorbs much of the harmful ultraviolet radiation streaming from the Sun (see OZONE LAYER). However, despite its importance, oxygen was not discovered until the late 18th century.

The honor of the discovery generally goes to English chemist Joseph Priestley (1733–1804). In 1774, he heated mercuric oxide (HgO) in a closed vessel by concentrating the Sun's rays through a lens. The mercuric oxide decomposed into free mercury and a gas, which Priestley found would support combustion better than air itself. He called it dephlogisticated air (see the box on page 1137). The gas had been discovered independently by Swedish chemist Carl Scheele (1742–1786) a year earlier, but he did not publish his findings until 1777. French chemist Antoine Lavoisier (1743–1794) gave the gas its present name—*oxygen* means "acid maker"—because he believed that all acids contained oxygen.

Oxygen in the air will keep this forest burning until all the combustible material has been consumed.

Physical and chemical properties

At room temperature and normal atmospheric pressure, oxygen is a colorless, odorless, and tasteless gas. Its atomic number is 8 and its atomic weight 15.9994. At 1 bar of atmospheric pressure, its boiling point is −297.35°F (−182.97°C), and its melting point is −361.82°F (−218.79°C). Although it is only slightly soluble in water—100 volumes of water will dissolve about 3 volumes of gas—the amount of oxygen dissolved in Earth's waters is essential to marine life.

Oxygen is a highly reactive element, with a strong attraction for the electrons of other elements and, in this respect, is second only to fluorine among all the elements (see ELEMENTS). For this reason, oxygen combines directly with nearly all elements, and oxides of the others (some of which are unstable) can be made by chemical reactions (see CHEMICAL REACTIONS). Most oxides form either spontaneously, when an element is exposed to oxygen—as when iron rusts in the presence of air and water or when the metal potassium bursts spontaneously into flame—or under the influence of heat or catalysts. Combustion of any substance is the oxidation of its component elements.

One of the most important reactions of oxygen occurs in living organisms at body temperature. This reaction is respiration, which liberates the energy required to sustain life. Although the chemical processes of respiration involve a number of complex reactions, it can be summarized as the oxidation of glucose to produce carbon dioxide, water, and energy:

$$C_6H_{12}O_6 + 6O_2 \rightarrow 6CO_2 + 6H_2O + \text{energy}$$

Without oxygen to support respiration, most animal life could not survive.

Oxides

The combination of one or more atoms of oxygen with one or more atoms of another element produces an oxide. The characteristics of such oxides are determined not only by the identity of their atoms but by the nature of the bonds that link them together (see CHEMICAL BONDS). For example, the bonds of metallic oxides tend to be ionic. An example is barium oxide, which dissolves in water to form barium hydroxide. Metallic oxides that react in this way are called basic oxides because they form basic solutions in water.

CORE FACTS

- Oxygen is the most abundant element in Earth's crust and makes up nearly 21 percent of the atmosphere.
- Oxygen is essential for respiration in almost all forms of animal life.
- Oxides can be acidic, basic, or amphoteric.
- The oxygen cycle relies upon photosynthesis in plants to maintain sufficient oxygen in the atmosphere.

CONNECTIONS

- The **OZONE LAYER** in the upper **ATMOSPHERE** absorbs much of the **ULTRAVIOLET RADIATION** from the **SUN**.

Symbol: O
Atomic number: 8
Atomic mass: 15.9994
Isotopes:
16 (99.76 percent),
17 (0.048 percent),
18 (0.20 percent)
Electronic shell structure: $[He]2s^22p^4$

THE OXYGEN CYCLE

Earth is a closed system: as free oxygen is converted to compounds by one set of processes, this combined oxygen is converted back to free oxygen by another set of processes. Animals remove oxygen from the air in the process of respiration and, in turn, release the oxygen-containing compounds carbon dioxide and water. Photosynthetic organisms, such as green plants and algae, use the carbon dioxide and water to make glucose, and then release the oxygen, which is recycled into the atmosphere. (Before the evolution of green plants during the Paleozoic era, free oxygen was virtually absent from the atmosphere.)

Oxygen is recycled continuously in Earth's closed system by many different processes. Free oxygen becomes combined oxygen, which is then converted back to free oxygen.

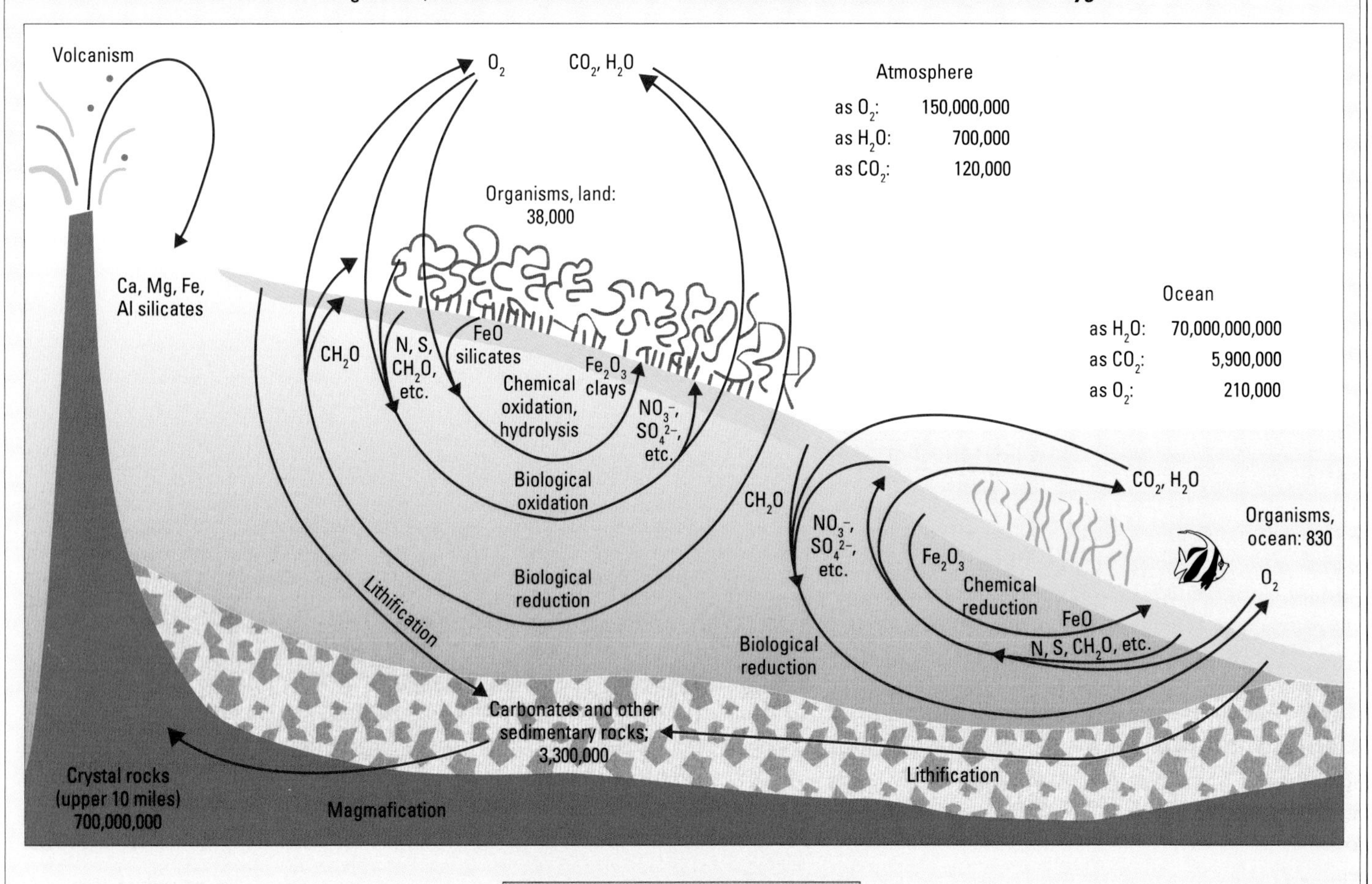

A Closer Look

The atoms of nonmetallic oxides form covalent bonds and tend to produce acidic solutions in water. For example, the gas sulfur dioxide dissolves in water to produce sulfurous acid. This reaction is the first step in the production of acid rain. Sulfur dioxide enters the atmosphere as a by-product of the burning of fossil fuels, most notably sulfur-rich coal. There it combines with water vapor to produce sulfurous acid, which, when exposed to sunlight and atmospheric oxygen, is converted to sulfuric acid.

Some oxides are neither basic nor acidic. Such oxides are called amphoteric, from the Greek meaning "both." They do not dissolve readily in water but will dissolve in either a strong basic or acidic solution. Some metals form oxides that can be acidic, basic, or amphoteric, depending on the oxidation state of the metal. For example, chromium can form CrO (oxidation state +2), Cr_2O_3 (+3), and CrO_3 (+6). The first oxide is basic, the second is amphoteric, and the third is acidic.

Highly reactive metals such as cesium, rubidium, and potassium react with oxygen to form a family of compounds called superoxides (CsO_2, RbO_2, and KO_2). Superoxides react with water to produce free oxygen. This property has been exploited in the design of masks used to supply oxygen for rescue workers. The masks contain a supply of KO_2, which reacts with the water vapor exhaled by the worker to produce oxygen for breathing:

$$2KO_2 + 2H_2O \rightarrow 2KOH + O_2 + H_2O_2$$

Slightly less reactive metals, notably sodium, calcium, strontium, and barium, react with oxygen to form a family of compounds called peroxides. When dissolved in water, peroxides do not produce oxygen but hydrogen peroxide (H_2O_2):

$$Na_2O_2 + 2H_2O \rightarrow 2NaOH + H_2O_2$$

Other metals and nonmetals react with oxygen to produce normal oxides. Some metals, such as barium, will react with oxygen to produce either a normal oxide, BaO (oxidation state +2), or a peroxide, BaO_2 (oxidation state +1).

Combustion

A very important chemical property of oxygen is its ability to support combustion, with the consequent release of energy, usually in the form of heat. This energy is produced mostly by the combustion of carbon-containing fossil fuels (hydrocarbons) such as coal, oil, natural gas, and gasoline. The complete combustion of a hydrocarbon produces heat, carbon dioxide, and water, for example:

$$CH_3CH_2CH_3 \text{ (propane)} + 5O_2 \rightarrow 3CO_2 + 4H_2O$$

There is usually a variety of other by-products, owing to incomplete combustion (if there is not enough oxygen) or to the presence of contaminants in the fuel.

The oxidation of organic compounds that already contain some oxygen can lead to the formation of many highly useful substances. For example, the controlled combustion of isopropyl alcohol produces the important industrial solvent acetone:

$$2CH_3CH(OH)CH_3 + O_2 \rightarrow 2CH_3COCH_3 + 2H_2O$$

Isolating oxygen

Oxygen is produced industrially by the liquefication and fractional distillation of air. The air is first freed from carbon dioxide, using lime or sodium hydroxide (caustic soda), and compressed to a pressure of about 30 bar. The air then expands through a nozzle and thus cools, and this cold air cools the pressurized air until it liquefies at about −321°F (−196°C).

At this stage, the liquid air is principally nitrogen (boiling point −320.4°F, or −195.8°C), argon (boiling point −302.4°F, or −185.8°C), and oxygen (boiling point −297.35°F, or −182.97°C). It then flows down a tower fitted with numerous plates, where it begins to evaporate. Nitrogen and argon, having lower boiling points, evaporate more easily than the oxygen, so the liquid air that reaches the bottom of the tower is made up of mainly oxygen (see PHASE TRANSITIONS). As it evaporates and passes upward through the liquid coming down, it becomes even richer in oxygen. After some time, 99 percent pure liquid oxygen accumulates at the bottom of the tower.

In the laboratory, oxygen can be prepared by several chemical reactions. The most common of these (see the diagram on page 1138) involves heating potas-

English chemist Joseph Priestley is often given credit for the isolation of oxygen, although he did not recognize its precise role in combustion.

OXYGEN AND THE PHLOGISTON THEORY

As scientists began to study the nature of matter during the 17th century, they were puzzled by combustion. When substances were burned or often merely heated, they changed into something different. Why? German professor of medicine George Ernst Stahl (1620–1734) proposed that the "fire" that the ancient Greeks thought was one of the four elements was a substance he named phlogiston (meaning "fiery matter"). This substance, he said, was present in all combustible matter and was driven out by heat. This theory would explain how metals were obtained from ores by smelting: litharge, for example, was lead that had lost its phlogiston. When it was heated with charcoal—which was combustible and full of phlogiston—the phlogiston was returned to the "dephlogisticated" lead.

A substance that burned completely was considered to contain a great deal of phlogiston. Substances such as the metals iron and copper, which were very difficult to burn, were considered to contain only a little. The role played by air in combustion was not understood.

It was not until French chemist Antoine Lavoisier (1743–1794), late in the 18th century, developed accurate methods of weighing the products of combustion that the process of combustion was more completely understood (see LAVOISIER, ANTOINE). By collecting all the gases produced by the combustion, he showed that when sulfur and phosphorus burned in air, they formed substances that weighed more than the original gases. When Lavoisier heard of Priestley's experiment, he repeated it and pointed out that the mercury produced at the end weighed less than the original mercuric oxide. If it had taken phlogiston out of the air, then phlogiston must weigh less than nothing. Clearly there had to be something wrong with the phlogiston theory.

Substances that burned in air, on the other hand, increased in weight. Lavoisier reasoned that this increase was due to their combining with something in the air. Lavoisier dubbed this substance "eminently respirable air," because birds lived longer in it. In 1777, he renamed it oxygen. The phlogiston theory had been proved false.

HISTORY OF SCIENCE

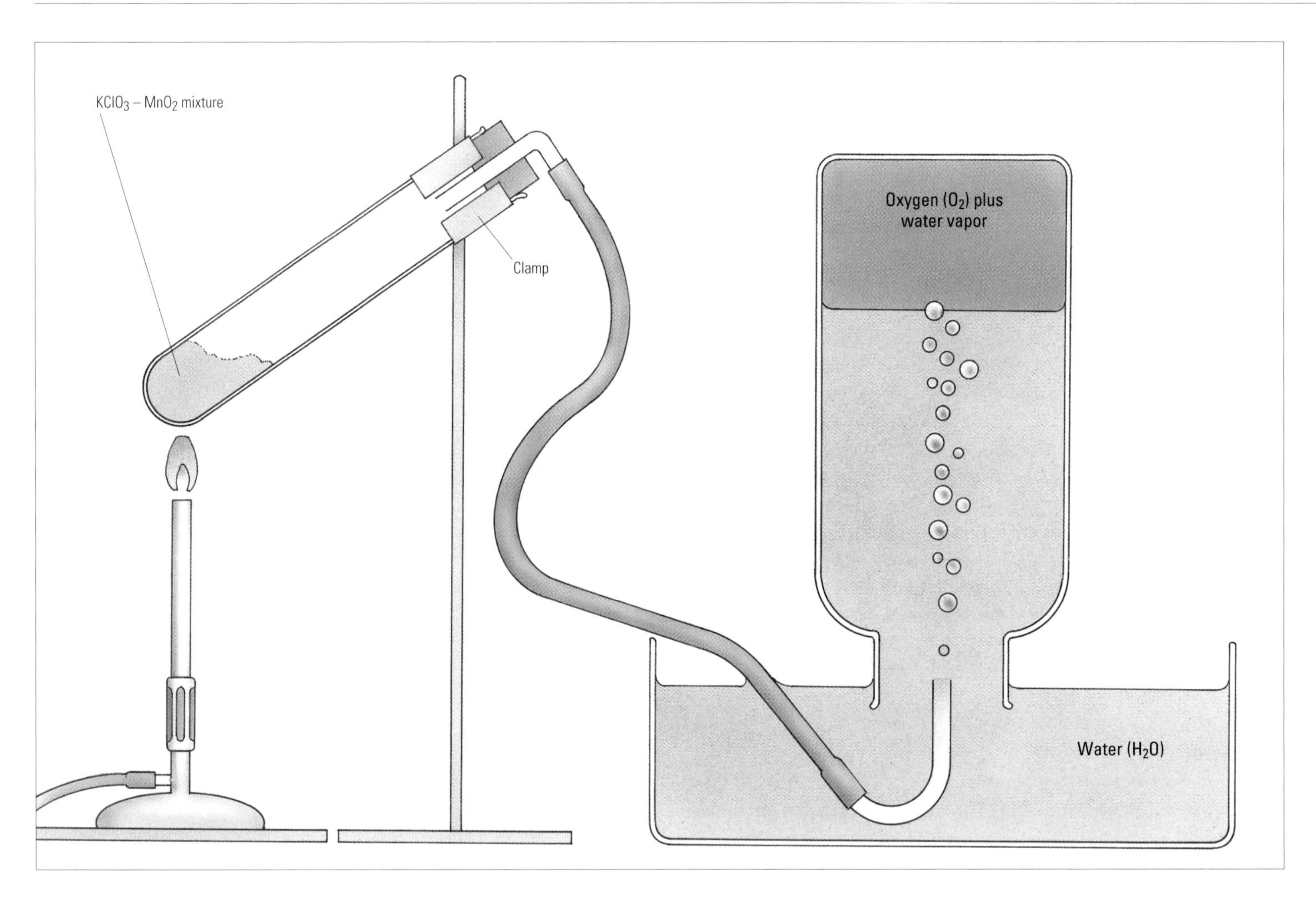

Oxygen can be prepared in the laboratory by heating potassium chlorate with manganese dioxide, which acts as a catalyst, and then cooling the mixture by passing it through water.

sium chlorate with a catalyst (see CATALYSTS), manganese dioxide, the end products being potassium chloride and oxygen:

$$2KClO_3 \ (+ \ MnO_2) \rightarrow 2KCl + 3O_2 \ (+ \ MnO_2)$$

Uses of oxygen

Industrially manufactured oxygen has many uses. In medicine, oxygen is used in respirators to aid the breathing of patients whose respiratory systems are impaired. Oxygen is also administered during major surgery to reduce the work of the heart and lungs. A special mixture of oxygen and other gases is used by deep-sea divers, military pilots, and astronauts.

In metal refining and steel manufacture, impurities such as sulfur, carbon, and phosphorus are removed by oxidizing them by injecting pure oxygen into the molten metal. Oxygen is used in oxyacetylene torches to produce the high temperatures needed to weld or cut metals. In this case, acetylene (ethyne) is the fuel, and oxygen is the oxidizer.

Oxygen is also used in the synthesis of many industrial chemicals or their precursors. For example, using controlled oxidation, natural gas, which consists mostly of methane (CH_4), yields chemicals such as acetylene (C_2H_2), ethylene (C_2H_4), and propylene (C_3H_6), which can be used as building blocks to synthesize a wide variety of more-complex organic compounds. Finally, a very powerful explosive can be made by impregnating charcoal with liquid oxygen just before use.

C. PROUJAN

See also: CATALYSTS; CHEMICAL BONDS; CHEMICAL REACTIONS; ELEMENTS; LAVOISIER, ANTOINE.

OXIDE TREASURES IN THE EARTH

Minerals are potentially valuable natural substances found in the earth. Some are ores; some are the stuff of precious gems; many are oxides. The table below lists some important mineral oxides.

Mineral name	Formula
Bismuth ocher	Bi_2O_3
Cassiterite	SnO_2
Chromite	$FeCr_2O_4$
Corundum (occurs as ruby, sapphire)	Al_2O_3
Cuprite	Cu_2O
Hematite	Fe_2O_3
Magnetite	Fe_3O_4
Pitchblende	UO_2
Pyrolusite	MnO_2
Senarmontite	Sb_2O_3
Zincite	ZnO

Further reading:

Cobb, Cathy, and Harold Goldwhite. 2002 *Creations of Fire: Chemistry's Lively History from Alchemy to the Atomic Age.* Cambridge, Mass.: Perseus Publishing.

Emsley, John. 2001. *Nature's Building Blocks: An A–Z Guide to the Elements.* New York: Oxford University Press.

OZONE LAYER

The ozone layer is the region of the upper atmosphere where most atmospheric ozone is found

The ozone layer is the layer of air containing ozone gas that is found high up in Earth's atmosphere, some 9 to 22 miles (15 to 35 km) above Earth's surface. This thin layer encircles the entire globe and acts as a filter, cutting out much of the harmful ultraviolet (UV) radiation that streams constantly from the Sun (see ULTRAVIOLET RADIATION). It consists of a mixture of gases, including free oxygen atoms, oxygen molecules, and ozone. The proportions of each fluctuate with the seasons, mainly in response to changes in the intensity of solar radiation and temperature.

Ozone is a triatomic molecular form of oxygen, with the chemical formula O_3. In the air people breathe, oxygen generally occurs in the less reactive diatomic form, O_2. Ozone is a bluish, reactive gas, and it has a sharp smell. Although ozone and oxygen are made of the same atoms, ozone's chemical properties are different from those of oxygen. While oxygen is essential for respiration, with much of the living world dependent upon it, ozone is poisonous to most forms of life, with less than one molecule per million in air being toxic to humans.

At ground level, ozone is a poisonous pollutant (see POLLUTION). A constituent of urban smog, it can worsen respiratory problems such as bronchitis and asthma. This ground-level ozone is formed when hydrocarbons react with oxides of nitrogen (both products of pollution) in sunlight (see HYDROCARBONS; NITROGEN AND NITROGEN CYCLE). Ozone is also formed naturally by lightning, which has enough energy to split O_2 molecules.

As well as damaging human health, ozone stunts the healthy growth of plants. Major cities such as Beijing, London, Mexico City, and Los Angeles, and their surrounding areas, all suffer from the effects of ozone from time to time.

Ozone formation and breakdown

The ozone layer is formed from oxygen in the upper atmosphere in the presence of solar radiation. A type of UV radiation provides the energy to split molecules of oxygen into their constituent atoms, some of which then recombine to form ozone, which itself is a good absorber of UV radiation.

The rates of formation and breakdown of ozone are more or less balanced in the ozone layer, and this balance seems to have remained stable over long periods of geological time. In October 1982, however, scientists working in the Antarctic were taking some routine measurements of the upper atmosphere and found, to their surprise, that there seemed to be a hole in the ozone layer—a patch in the normally uniform layer that contained less ozone and through which intense UV radiation was reaching the surface.

Similar measurements were taken in the next two seasons to check the equipment on the ground and for comparison with satellite readings from above the layer. The measurements grew larger not only confirmed the original findings but indicated that the hole, which appeared each southern spring, grew larger each year. The results caused widespread concern, and the hunt for the cause began in earnest (see MOLINA, MARIO). In the following years, the ozone hole grew larger, and in 1991 it was 13 times its size in 1982. Human activities were affecting the balance in the ozone layer and causing a net destruction.

It is a similar story in the Northern Hemisphere. Recent measurements are beginning to detect a similar depletion appearing over the Arctic. The ozone layer over southern Canada has thinned by an average of about 6 percent a year since the late 1970s, when human activities first began to affect the ozone. Levels in Scandinavia, Greenland, and Siberia reached an unprecedented 45 percent depletion in 1996.

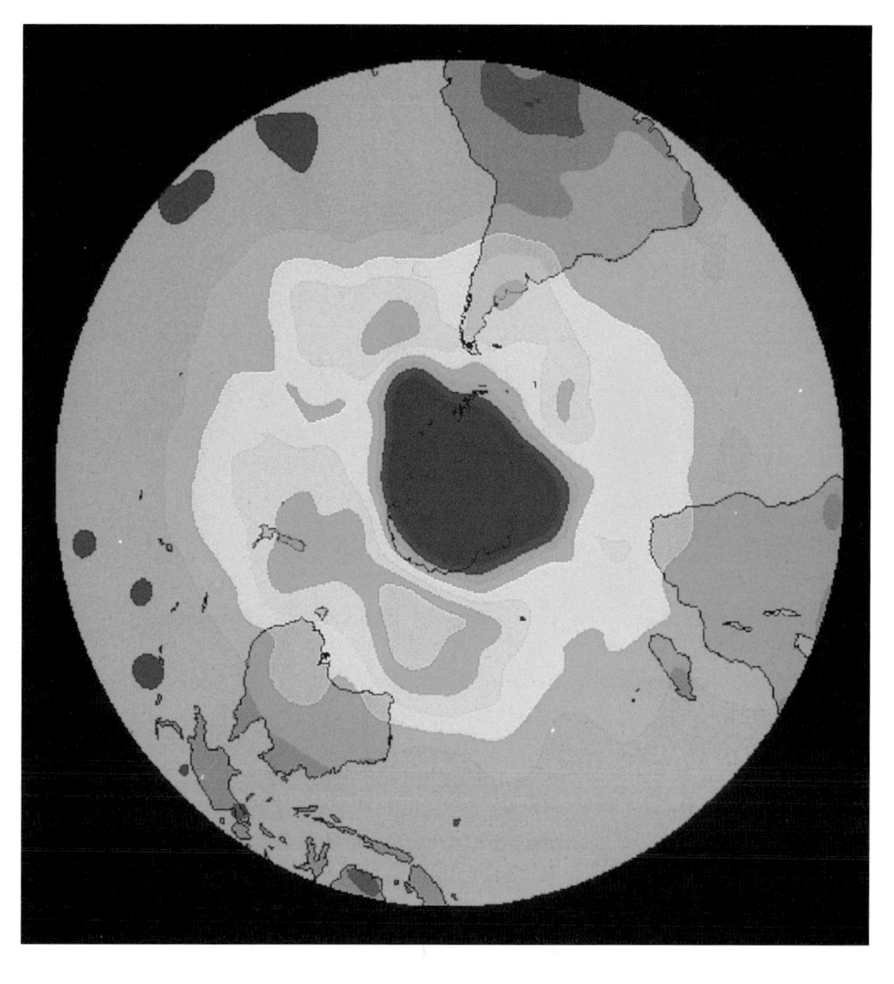

The image above shows the total atmospheric ozone concentration in the Southern Hemisphere. The Antarctic ozone hole is at the center, colored light blue, dark blue, and purple.

CORE FACTS

- The ozone layer is mainly found in the lower part of the stratosphere and contains the majority of atmospheric ozone (approximately 97 percent).
- Scientists believe that human-made chemicals are largely responsible for the recent destruction of the ozone layer.

CONNECTIONS

- Ozone **GAS** is the triatomic molecular form of **OXYGEN**.
- The ozone layer is found in the layer of the **ATMOSPHERE** called the **STRATOSPHERE**.

Aerosol sprays used to be powered by chemicals called chlorofluorocarbons, which, when released into the atmosphere, are believed to be a major contributory factor in the destruction of the ozone layer.

The effects of ozone layer thinning

It has been known for some time that the UV component of even normal levels of solar radiation can be harmful. UV has been shown, for example, to be a major cause of skin cancer, a disease that kills many thousands of people every year. UV also can cause cataracts and blindness, as well as a general lowering of the effectiveness of the immune system. In recent years, there have been worrisome increases in cataracts and skin complaints in those regions that might expect to be affected by increased UV following ozone layer thinning.

Plants, including crops, may also be damaged by excess UV radiation, so food supplies may suffer as well. There may also be effects in the sea, where excess UV may destroy plankton. Phytoplankton absorb carbon dioxide, one of the so-called greenhouse gases, so this effect could result in a buildup of carbon dioxide in the atmosphere, which could contribute to global warming (see GLOBAL WARMING).

Most scientists now agree that the most likely cause for the destruction of atmospheric ozone is the effect of certain humanmade chemicals, principally the chlorofluorocarbons (CFCs; see HALOGENS). CFCs have been used widely, notably in refrigerators in aerosol sprays, in foams, and in cleaning fluids. Several other chemicals also react with ozone, including carbon tetrachloride and methyl chloroform, both used in solvents, and halon dichlorodifluoromethane, used in fire extinguishers.

CFCs and other chemicals gradually drift upward into the higher levels of the atmosphere, where they accumulate before eventually breaking down under the intense radiation levels. On breaking down, they release chlorine, which reacts with ozone and converts it to oxygen. The various forms of CFCs are responsible for about 80 percent of ozone depletion.

In 1987 an international treaty, the Montreal Protocol, agreed to cut the use of CFCs by half by 1996, and a later meeting agreed to phase them out altogether by 2000. It is estimated that the ozone layer should recover by about 2050 if all humanmade ozone-depleting substances are eliminated. However, long-term predictions are uncertain because the processes of ozone depletion are not understood.

M. WALTERS

See also: GLOBAL WARMING; HALOGENS; HYDROCARBONS; MOLINA, MARIO; ULTRAVIOLET RADIATION.

Further reading:

Christie, Maureen. 2001. *Ozone Layer*. New York: Cambridge University Press.

OZONE: "TO SMELL"

The characteristic smell of ozone, produced by an electric spark, was noticed as long ago as 1785 when scientists first began experimenting with the newly discovered voltaic battery (see BATTERIES). In 1840 German chemist Christian Schönbein (1799–1868) identified ozone as a new gas and gave it its name, from the Greek meaning "to smell." Its chemical formula was established in 1866.

For many years, people attributed the fresh, sharp, smell of seaside air to a higher concentration of ozone, but there is no evidence for this opinion, and normal air at Earth's surface is calculated to contain about one molecules of ozone per 100 million molecules of other air gases.

Pure ozone has been separated from liquefied air. It is a dark blue, very explosive liquid, with a boiling point of –170°F (–112°C). It rapidly decomposes into oxygen. Ozone is a very powerful oxidizing agent, and it will react with many substances in conditions where oxygen will not, generally giving off oxygen at the same time. All metals, except platinum and gold, are oxidized, as are many organic substances, including rubber and cellulose. Organic substances containing double bonds, such as benzene and turpentine, will absorb ozone completely, forming ozonides, with an attached ring structure containing three oxygen atoms. Ozone has disinfectant properties and is sometimes used in swimming pools.

A CLOSER LOOK

PACIFIC OCEAN

The Pacific Ocean lies between Asia and Australia to the west and North and South America to the east

The Pacific is the oldest and largest of Earth's oceans, covering an area nearly as large as the combined total of the other three: the Atlantic, the Indian, and the Arctic Oceans. Encompassing one-third of the surface of Earth, the Pacific has an area greater than all the landmasses combined; its volume is more than double that of the Atlantic, the second-largest ocean.

Facts and figures

Including the seas that cluster along its western edge, the Pacific Ocean covers 60,300,000 square miles (179,700,000 km^2). From the Bering Sea in the north, it is 9,600 miles (15,450 km) to Antarctica. From Malaysia in the west, it is 12,000 miles (19,300 km) to the coast of South America.

Of all the oceans on Earth, the Pacific has the greatest depths. The Mariana Trench, which runs approximately north to south 1,550 miles (2,500 km) east of the Philippines and 1,250 miles (2,000 km) north of Papua New Guinea, at 36,198 feet (11,033 m) deep could swallow up Mount Everest, Earth's tallest mountain, with more than a mile left to spare. The average depth of the Pacific is 13,216 feet (4,028 m).

This basin is fed by the waters from numerous rivers. From North America flow the Yukon, Fraser, Columbia, Sacramento, and Colorado. From Asia flow the Amur, Yellow, Yangtze, Hsi, and Mekong into a string of adjoining seas, including the Bering, Okhotsk, Japan, Yellow, East China, South China, Philippine, Sulu, Celebes, Molucca, Java, Banda, Arafura, Timor, Coral, Tasman, and Ross. With mountains shielding it from the greatest expanses of the southern continents, the Pacific receives only minor river flows from Australia and South America.

The Pacific is an ocean of variety, from the humid waters of the South Pacific to the ice-choked waters of Antarctica and from the warm trade winds enjoyed by mariners near the equator to the dreaded williwaws (sudden violent gusts of cold air), where the warm Pacific meets the ice-cold Bering Sea.

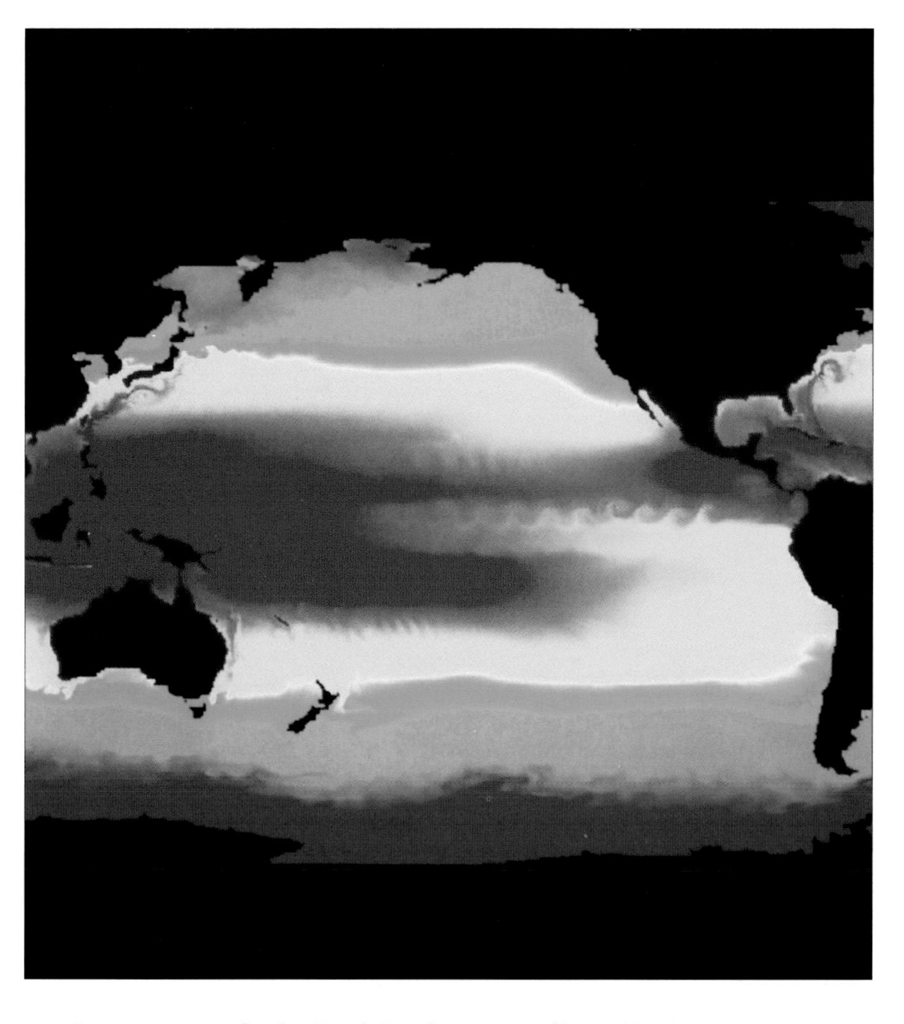

This computer-generated image shows sea-surface temperature distribution in the Pacific Ocean. The temperatures are color coded from red (warmest) through green and yellow to blue (coolest).

Contrary to myth, the South Pacific is generally cooler than the North Pacific, kept cool by the influence of Antarctica. The North Pacific is surrounded by a larger landmass that keeps ocean temperatures relatively warm.

Ocean currents

Like the general pattern of Pacific winds, the major currents of the ocean flow clockwise north of the equator and counterclockwise to the south. The nine major currents of the Pacific include the West Australian, Peru, and South Equatorial in the South Pacific; the Kuroshio (Japan), California, and North Equatorial in the North Pacific; the North Pacific–Alaska Current eddying westward in the Gulf of Alaska; the Equatorial Countercurrent flowing eastward between the North and South equatorial currents; and the Antarctic Circumpolar Current—Earth's most powerful—circling the globe just north of the frozen continent.

Pacific islands and seafloor topography

The western Pacific seafloor features a broad continental shelf with numerous lengthy, narrow arcs of islands: Aleutian, Kuril, Ryukyu (including Japan), Micronesia (including the Carolines, Kiribati, the

CORE FACTS

- The Pacific is the largest of Earth's oceans, covering an area of approximately 64 million square miles (167 million km^2).
- The average depth of the Pacific Ocean is 14,040 feet (4280 m).
- Like the general pattern of Pacific winds, the major currents of the ocean flow clockwise north of the equator and counterclockwise to the south.
- The seafloor is littered in some areas with nodules containing high concentrations of cobalt, copper, iron, nickel, and manganese.

CONNECTIONS

- **OCEANIA** consists of a large number of coral and volcanic **ISLANDS** scattered across the vast Pacific Ocean.
- **EL NIÑO** is an important **CURRENT** of the Pacific.

The island chain of Hawaii is volcanic in origin. The islands are of varying age, and some, such as the Big Island, are still volcanically active. A new Hawaiian island, Loihi, is just starting to form. It may take 50,000 years for it to surface.

Marianas, the Marshalls, and the Ellices), Melanesia (the Bismarcks, Fiji, New Caledonia, the Solomons, and New Hebrides), Papua New Guinea, Tonga, and New Zealand. These islands were formed as magma, welling up from beneath the Pacific plate, melted the land to a depth of about 60 miles (100 km). Some islands split off the continent, and new volcanoes appeared (see ISLANDS; OCEANIA; VOLCANOES). On the ocean side of these chains, the Pacific plate plunges far under the continental plates forming spectacularly deep trenches.

FROM WHERE DID THE WATER COME?

Water has probably existed on Earth almost as long as the planet has existed as a solid body, but from where did the water come? The ocean must have been created between 4.6 billion years ago (when Earth formed; see EARTH, PLANET), and 3.8 billion years ago, when the oldest known rocks were formed (see ROCKS). Geologists are not entirely sure, however, how the ocean formed. The most probable explanation is that water condensed from water vapor in Earth's ancient atmosphere. The water vapor and other gases were released from Earth's interior by volcanoes and other eruptions of magma.

Water is slowly removed from the hydrosphere by being buried along with sediments as they are slowly transformed into sedimentary rocks . Water also can react with rocks and become incorporated into the chemical composition of their minerals. At the same time, water continues to be added to the hydrosphere through volcanic activity. It is difficult to say whether the subtraction and addition of water are exactly equal, but the difference is likely to be so small that people can assume the hydrosphere has essentially reached a state of balance.

A CLOSER LOOK

In the east, the continental shelf is narrow along the American continents (see CONTINENTAL SHELVES). The North American plate has overridden the northern reaches of the Pacific's mid-ocean ridge—the East Pacific Ridge—but its presence is still very much evident. It coincides with the Gulf of California and the fault system that includes California's infamous San Andreas fault (see FAULTS). Still farther north, the Cascade Mountains show, by their continuing volcanic activity, the forces that lie deep below.

The central Pacific Ocean is sprinkled with the island groups that make up Polynesia, including the Cook Islands, the islands of the Hawaiian archipelago, the Marquesas, Samoa, the Society Islands, Tonga, and the Tuamotus. Many of these islands and especially the Hawaiian archipelago are forming as a result of hot spots and mantle plumes. These plumes of magma originate deep in Earth's mantle and appear to be independent of plate movements (see MAGMA). The Big Island of Hawaii is the best example of a large shield volcano forming above a still-rising tail of a mantle plume beneath the oceanic lithosphere. Older Hawaiian islands to the north and west are part of the oceanic plate that has moved away from direct contact with this plume.

In the central Pacific, the abyssal plain has the most regular features anywhere in the ocean. The North Pacific Deep, more than 18,000 feet (5,500 m) deep, covers more than 1 million square miles (2.6 million km^2). In the deep waters above the abyssal plains, chemical processes create metal-bearing coatings around objects such as fish bones. These nodules, containing high concentrations of cobalt, copper, iron, nickel, and manganese, litter extensive areas of the Pacific. The origin of manganese nodules is not clearly understood. The metals may be derived from magmatic fluids issuing onto the seabed and developing around a nucleus.

Life at the bottom

The Pacific Ocean contains a wide range of plants and animals, many more than are found in the world's other oceans. The animals that live there have to survive tremendous water pressures and live with very little, if any, light. Most animals in the deeper parts of the ocean feed on organic detritus, which sinks from the surface. However, animals that live at volcanic rifts (see HYDROTHERMAL VENTS; VOLCANOES) are dependent on bacteria that obtain energy from sulfur compounds rather than the Sun.

J. RHODES

See also: ANTARCTICA; ARCTIC OCEAN; ATLANTIC OCEAN; CONTINENTAL SHELVES; EARTH, PLANET; FAULTS; HYDROTHERMAL VENTS; INDIAN OCEAN; MAGMA; OCEANS AND OCEANOGRAPHY; PLATE TECTONICS; ROCKS; VOLCANOES.

Further reading:

Severdrup, K. A., A. C. Duxbury, and A. B. Duxbury. 2003. *An Introduction to the World's Oceans.* New York: McGraw-Hill.

PAINTS AND PIGMENTS

Paint is a decorative and protective liquid coating consisting of a pigment suspended in a solvent and binder

Australian aboriginal rock art proves that paints have been used for many thousands of years. Prehistoric rock drawings have also been found in caves in France and Spain.

Paints have been used for decoration since prehistoric times. Paintings dating from around 15,000 B.C.E. have been discovered in caves in France and Spain, where early humans drew the animals they hunted and colored them with minerals they found close by. Ancient Egyptians developed a much wider range of colors and used dyes such as indigo to make blue and red pigments.

A paint has three basic constituents: the coloring matter, which is a dye or pigment (a dye dissolves in the paint system, a pigment does not; see DYES); a liquid solvent; and a binding substance that will "fix" the paint so that it is not removed easily. The latter two constituents are called the vehicle. In principle, basic paint technology has not changed over thousands of years, but during the 20th century the individual constituents have been developed so much that modern paints are very different from traditional materials.

CORE FACTS

- Paints are composed of pigments dispersed in a vehicle.
- Vehicles are a mixture of resins and solvents or thinners.
- The oldest pigments were natural minerals, and many of these minerals are still used in modern paints.
- Resins dry by oxidation or polymerization to form a hard but flexible protective coating.

Paint vehicles

Analysis of ancient paintings has shown that the usual vehicles were gum arabic, egg white, gelatin, casein (from milk), starch, or beeswax. Some of these were diluted with water, but they dried to form an insoluble coat that has resisted the effects of humidity and other atmospheric factors for many centuries. In Europe, up to the 14th century, artists traditionally used egg white as the vehicle. One of the major decorative techniques was fresco painting (from the Italian meaning "fresh") on newly applied plaster.

Gradually, oil painting was adopted, in which the vehicle was oil or a mixture of turpentine and linseed. Turpentine was used as a thinner for the thick oil, and it slowly evaporated; the oil oxidized under the action of air and formed an insoluble solid coating (see LIPIDS). Because this coating became dull in time, various varnishes were added to the paint or applied afterward to maintain the gloss and enhance the depth of color. The choice of different oils and derivatives of oils widened the scope of application.

Pigments

Until the 20th century, nearly all pigments were mineral in origin. Ancient artists used red and yellow ochers, which are varieties of iron oxide; blue and green oxides of copper or the rarer gemstone lapis lazuli (see the picture on page 1145); red cinnabar, or

CONNECTIONS

- The coloring matter in paints consists of either pigments, which do not dissolve in the paint system, or **DYES**, which do dissolve.
- Synthetic resins are made by modifying **OIL** or fatty acids with **CARBOXYLIC COMPOUNDS** and **ALCOHOLS**.

mercuric sulfide (see MERCURY, METAL); white lead carbonate; and charcoal or carbon forms. Plant and animal pigments were also used, but many of these substances were fugitive—that is, they faded in time—or migratory—they diffused through the painting.

HOW PAINTS ARE TESTED

A paint is of little use if it fails. Weather and other environmental effects can have a big impact on a paint's durability. Tests to mimic these effects measure a paint's performance. There are two different types of tests: those that measure the paint's physical performance and those that look at how they resist chemical attack (the picture below shows peeling paint).

For example, hiding power is tested by painting different thicknesses over printed charts with black and white areas. The thickness at which the black and white become indistinguishable is a measure of the hiding power.

Adhesion is tested by cutting a lattice pattern into the layer of paint, then covering it with adhesive tape, rubbing it, and removing the tape. Hardness can be measured by penetrating the paint film with a needle under known pressure. Flexibility and resistance to abrasion are measured. Color and gloss are also determined. Tests for performance against chemical attack include measuring water resistance, since paint immersed in water tends to blister. Finding out how paints are affected by the detergents and alkalis found in cleaning fluids is also important in paint technology.

Finally, paints must be tested for fading. In the laboratory, they are exposed to an ultraviolet lamp. Samples are placed outdoors, both to see how they respond to lengthy periods of sunlight and to observe their resistance to the weather.

A CLOSER LOOK

One of the skills of the traditional paint makers was the grinding of these pigments to extreme fineness so that the particles could be dispersed evenly through the vehicle, which would develop the full color strength and not form visible lumps in the finished work. As a result, the paint was often nearly transparent. With the development of oil-based vehicles, paints that could be applied to almost any surface, the surface was usually primed with white, which added brightness and luster to the more transparent colors that were laid over it.

Artists generally primed canvas or wood with paint made of gesso (plaster of paris) or finely ground chalk; paints for domestic or industrial use were applied over an undercoat of white lead carbonate. Where white paint was required, particularly in the interior and on the exterior of homes, this "white lead" was usually used as the finish. However, the growing use of coal during the 19th century released large quantities of hydrogen sulfide into the atmosphere. This gas converted the lead carbonate in the paint into black lead sulfide, and so it became necessary to develop other white pigments. Also, lead poisoning became a major problem, particularly in children (see LEAD).

Modern paint technology

For some 500 years paint remained an expensive luxury: it was manufactured by skilled craftspeople and could be afforded only by the rich. However, as iron machinery, bridges, and other structures were developed in the 18th and 19th centuries during the industrial revolution, it became necessary to protect them against rusting. The demand for paint grew, and gradually factories were set up for industrial paint manufacture. It was not until the early 20th century that major advances in paint making were made.

Paints have two main purposes: decoration and protection. A successful paint must remain attached to the surface it is protecting. Different surfaces have different requirements, so a paint for metal will not have the same formulation as one for wood. A modern paint will contain a range of different substances: pigments and extender pigments, solvents (or thinners), resins (binders), and plasticizers.

Pigments are the part of a paint most obvious to the naked eye because they give it its color. They also impart hiding power, because they conceal what is underneath the layer of paint. Some pigments have a particularly high hiding power, so they can be supplemented by extender pigments—cheaper substances that have the same particle size as the main pigment. Many pigments also have other properties in addition to their color. They include, perhaps most notably, those providing an anticorrosive action, vital for protecting metal, and biocidal pigments, which are used to prevent the fouling of ships' hulls with barnacles.

The resin in paint is used to make the film (which binds the pigment to the surface) and provide protection. The oldest form of resins are the oleoresinous varnishes, made by cooking resins such as wood resin with linseed or other drying oil. The drawbacks of these formulations were that they dried slowly and did not cover particularly evenly. Oil-based paints will usually lose their high-gloss finish within a couple of years, and because they are not very hard, they have only limited weather resistance.

Much more widely used nowadays are alkyd resins. They are polyester compounds, produced by modifying oils or fatty acids with polyfunctional (with more than one functional group) acids, such as phthalic, maleic, or sebacic acids (see CARBOXYLIC COMPOUNDS), and polyfunctional alcohols, such as glycerol, pentaerythritol, or ethylene glycol (see ALCOHOLS AND ETHERS). Other resins used in paints include polyurethanes, silicones, epoxy resins, and polyacrylates. A similar class of paints are the cellulose lacquers, based on nitrocellulose or less flammable substances such as cellulose acetate and propionate. These lacquers have a very high gloss and are used for such applications as automobile finishes.

A different type are the latex paints, in which partly polymerized vinyl acetate or acrylic esters are dispersed in water. These paints have many advantages, including low odor, absence of toxic or flammable fumes, and the ability to be applied to damp surfaces. They dry rapidly, forming stable polymer films, and are very durable. They have the disadvantage of forming films with poor gloss, the result being a matte or, at best, a semigloss finish.

Inorganic pigments

Pigments must be ground very fine so that they can be dispersed in the vehicle. The most important white pigment is titanium dioxide because of its extremely good hiding power. It is made from either of the two ores of titanium: rutile or ilmenite. Rutile is a natural form of titanium dioxide but is less widespread than ilmenite, which contains titanium and iron. There are two grades of titanium dioxide pigment: rutile and anatase, which have different crystal structures. Rutile is more suitable for decorative purposes and is used in emulsion and gloss paints. The anatase form gives good whiteness but is not very weather resistant.

Zinc oxide is otherwise known as the artist's pigment Chinese white. Not very widely used today, it is important in certain specific applications, such as in paints used in warm, damp climates, as an antifungal agent. Zinc sulfide is another white pigment, but titanium dioxide remains the most important.

Extenders, such as kaolin (china clay) and chalk, are essentially transparent in paints, so they do not add to either a paint's color or to its hiding power. They are cheaper than pigments, so incorporating a quantity into a paint reduces the cost. This is not the only reason for their use—they also affect the paint's consistency and so are used to change its properties.

Many naturally occurring pigments contain iron. They have good hiding power and light fastness. Sienna pigments are brownish yellow and are mined mostly in Italy. They contain between 40 and 70 percent iron oxide with a small amount of manganese dioxide. The untreated mineral is called raw sienna; if it is heated to a high temperature, it becomes dehydrated and leaves burnt sienna, which is orange-red.

Lapis lazuli is a rare, deep-blue gemstone, formerly used as the source of the blue pigment ultramarine. It contains blue lazurite (sodium aluminum silicate), white calcite (calcium carbonate), and small amounts of other minerals.

Umbers are a much darker brown than siennas. They contain a smaller quantity of iron oxide and a much larger amount of manganese dioxide. Much is mined in Cyprus, the mined pigment being known as raw umber. Burnt umber is a rich red-brown. Ocher pigments are clays that also contain iron oxides. The colors vary enormously from one source to another, but most are shades of yellow. Spain is a good source of red ferric oxide pigments, which are a particularly bright shade of red.

Two compounds of cadmium provide brilliantly colored pigments that are very durable. Cadmium selenide is cadmium red, and cadmium sulfide is cadmium yellow. Cadmium yellow must not, however, be mixed in paint that contains lead compounds, since it will steadily darken with the formation of lead sulfide (see TRANSITION ELEMENTS).

Natural blue pigments are much rarer than yellows and reds. Ultramarine is an ancient pigment, originally made from the mineral lapis lazuli found in China and Iraq. It is now made synthetically by roasting a mixture of china clay and sodium sulfate with soda, carbon, and sulfur. It is bright blue; violets can be obtained by heating it in hydrogen chloride gas. Prussian blues, also known as iron blues, are synthetic pigments. They are iron hexacyanoferrate salts: the potassium salt $KFe[Fe(CN)_6]$ is used in paints and the ammonium salt $6Fe(NH)_2.Fe(CN)_6$, in printing inks.

The chromates of heavy metals are, in general, insoluble and often used as pigments. They are obtained from the brownish black mineral chromite and are an important group of yellow and greenish pigments. The most common black pigments are carbon blacks, which are made from oils or natural gas burned in a limited air supply to give carbon.

The actual shades of natural pigments vary greatly because of the differing amounts of impurities they contain. The colors of synthetic pigments are more reproducible. Synthetic red oxide, for example, is brighter than a mined source, because it contains almost 100 percent ferric oxide (Fe_2O_3).

Organic pigments

A much wider range of colors is found in organic pigments, although they are seldom as durable as inorganic pigments. (Almost no pigments are obtained today from natural sources, since the synthetic alternatives are more durable.) They can be divided into three classes: lakes, toners, and pigment dyestuffs.

Unlike pigments, dyes form a chemical bond with the material they are coloring. A lake is a dye that has been precipitated onto a white substance, such as alumina or china clay. In some lakes, the dye coordinates to the base and becomes an integral part of it. Toners are made by precipitating acid dyes with metal salts, usually barium, calcium, or manganese. Pigment dyestuffs include such classes of compounds as azo dyes and metal complex dyes.

Azo pigments are the most important chemical group of organic pigments. Like the azo dyes, their colors range mostly from yellow through orange to red. The yellow colors, in particular, are widely used, and they have replaced toxic lead chromate pigments in most paints. The arylamide pigment hansa yellow G, for example, is a bright red-yellow, and the arylamides range from red-yellow to green-yellow. Benzimidazol pigments are stronger colors that range from yellow to brown. Permanent orange G is the best known benzidine yellow; it is bright orange (see COLOR).

One of the most common blue pigments is copper phthalocyanine. It is a particularly stable compound, so it is widely used in paints, having to a great extent displaced Prussian blue. Green pigments are produced by preparing chlorine or bromine derivatives. Many other colors can be created by mixing pigments together. Pale pastel paints are made by mixing small quantities of colored pigments with titanium dioxide. Preferably, the different pigments should have compatible chemical properties so that the color of the paint is constant throughout its life.

Batches of ceramic pigments after kiln firing.

Thixotropic paints

The chance of painting oneself while painting the ceiling was significantly reduced with the introduction of thixotropic paints. Also called nondrip paints, they are in a gel form while still in the can, become more fluid when stress is applied (when the brush is dipped in and it is applied to the surface to be painted), and then become a gel again so that they do not run or sag. Substances that give nondrip properties include hydrogenated castor oils, montmorillonite clays, and aluminum alcoholates.

Manufacturing paints

Paints are manufactured by dispersing pigment particles in a liquid system to give a uniform consistency. In nonaqueous paints, the solid substances, such as pigments, are broken down by grinding them into uniform particle sizes. In this process, the finer particles are wetted by displacing the air or water present at the pigment surface with the paint medium. The dispersions are then stabilized to prevent the pigment from settling together in one mass—or flocculating (forming clusters)—or to prevent the viscosity from increasing on storage.

The manufacturing process for aqueous emulsion paints is slightly different. The pigment is dispersed in the aqueous phase and then mixed with the polymer emulsion. In white or pastel paints, the aqueous medium, in which rutile titanium dioxide is dispersed, is a dilute solution of sodium carboxymethylcellulose or cellulose ether or a dilute solution of an ammonium polyacrylate. Pigments in these paints often flocculate; an anionic surfactant in the formulation will prevent flocculation.

S. HOULTON

See also: COLOR; LIPIDS; MERCURY, METAL; TRANSITION ELEMENTS.

Further reading:

Zollinger, Heinrich. 2003. *Color Chemistry: Syntheses, Properties, and Applications of Organic Dyes and Pigments*. Weinheim, Germany: Wiley-VCH.

Herbst, W., and K. Hunger. 2004. *Industrial Organic Pigments: Production, Properties, Applications*. New York: VCH.

PALEOCLIMATOLOGY

Paleoclimatology is the scientific study of past climates over timescales from a few years to many millions of years

Paleoclimatology is the study of past climates. Understanding what has caused the cycles of climatic change in the past helps scientists make predictions about what is likely to happen in the future. The human impact on climate change—notably, the burning of fossil fuels and its likely impact on global warming (see GLOBAL WARMING)—can be placed in the context of natural climatic changes that have happened in the past and may be happening now.

Paleoclimatology is a truly interdisciplinary field of study. It involves, among other scientists, climatologists (who create models to describe, explain, and predict current climates and weather; see CLIMATE), archeologists (who study human artifacts from the past), and geologists (see GEOLOGY).

A volcano on the Siberian peninsula of Kamchatka spews ash and gases into the atmopshere. Throughout history volcanic eruptions have caused short-term climate changes.

Evidence for the nature of past climates

Paleoclimatologists can infer past climates from a wide range of evidence. Any preserved feature that is linked to climate and whose age can be determined can serve as a piece of the jigsaw puzzle in revealing the climate of the distant past.

In practice, the best sources of evidence for climatic changes over millions of years come from analyzing the composition of strata (layers) in sedimentary rocks, studying ancient organisms preserved as fossils, and examining chemical signatures locked in the buried rock (see FOSSIL RECORD; GEOCHEMISTRY; SEDIMENTARY ROCKS). Scientists reveal climatic changes over shorter timescales by analyzing gas bubbles trapped in ice, the growth rings of trees, and historical evidence.

Sediment layers

A sedimentary sequence has older layers lying beneath younger layers often in an orderly sequence. Detailed study of ancient sediments that are now lithified (turned to rock) can reveal the environmental conditions that existed when the sediments were deposited. For example, a layer of sandstone may infer desert conditions, a layer of coal suggests tropical conditions (see COAL AND COAL GAS), and deposits of rock fragments of varying sizes worn smooth, signify glacial conditions (see GLACIERS AND GALCIOLOGY; ICE AGES). Marine deposits that are rich in rock salt (halite) and gypsum (calcium sulfate) infer warm conditions with high rates of evaporation, while those laden with rocks of varying size suggest cool conditions when icebergs melted and deposited their load of rock fragments.

If sediments carry a strong magnetic record in the form of trapped iron oxide mineral grains, the data may be used to infer the latitudinal position (polar, temperate, or equatorial) when the sediments were deposited (see the box on page 1148).

CORE FACTS

- Among the approaches paleoclimatologists use to determine past climates, are the study of layers in sedimentary rock, fossil evidence, oxygen-isotope ratios in sedimentary deposits, tree ring data, and historical and archeological evidence.
- Scientists explain long-term climate trends, on timescales of millions of years, in terms of alterations in the distribution of landmasses, changes in levels of surface volcanic activity, mountain-building periods, and the removal of atmospheric carbon dioxide that becomes stored in carbon sinks beneath Earth's surface.
- Short-term climate changes, on timescales between a few years and many millennia, depend on the same factors that influence long-term changes, with the addition of sunspot activity cycles.

Fossil evidence

The distribution of remains of dead organisms (fossils) provide vital information about past climates. Almost any fossils can infer useful climatic information, but in practice, plankton, pollen, and insect remains are good sources of climatic data.

CONNECTIONS

- Scientific study of past climates sheds light on the degree of human influence in current **GLOBAL WARMING** trends.
- **SEDIMENTARY ROCKS** are key sources of evidence on which estimates of past climates are based.

MAGNETIC RECORDS

Earth behaves like a giant bar magnet, with magnetic poles near the North and South Poles (see MAGNETIC POLES). The angle that Earth's magnetic field makes with Earth's surface is steeper near the poles and shallower near the equator. If sediments contain fragments of iron that preserve the inclination of the magnetic field at the time they were formed, such readings can be interpreted to give the approximate latitude at which the sediments were formed.

A CLOSER LOOK

Different assemblages of animals, plants, and microbes flourish in different climates, and the succession of species in a sedimentary sequence can reveal the climatic record for that site. For example, the assemblage of microscopic plankton (floating organisms) living in the sea is characteristic for a given temperature regime—warm-water species are different from cool-water forms. The sequence of cool-water and warm-water plankton fossils in marine sedimentary rock reveals how the climate has changed in that region over thousands of years.

Scientists can reveal past climates on land through pollen analysis. Flowering plants produce pollen, tiny grains containing male sex cells enclosed a in a protective case. Under a microscope, the pollen grains of a given species are often distinctive, and because pollen grains are so hardy, they fossilize easily. Palynologists (pollen scientists) examine the sequence of pollen grains preserved in peat. The mix of pollen grains in a particular layer can be correlated with the climate those pollen-yielding trees prefer. The sequence of pollen mixtures in successive peat layers provides the record of climatic change over time. Surveys over a wide area reveal shifts in tree distributions. For example, the northward advance of spruce trees in North America has been tracked using pollen records. Since the peak of the last ice age 18,000 years ago, the northern limit of spruce distribution has shifted 1,300 miles (2,100 km) north in the eastern United States and Canada as the climate has warmed.

Paleontologists increasingly use fossil beetles as climate indicators (see PALEONTOLOGY). Beetles travel less far than pollen and so can be a better indicator of local climatic conditions. Computer databases that now document the body parts of thousands of species of fossil beetles enable paleontologists to correlate insect finds with climate.

Paleoclimatologists can follow the advance and retreat of glaciers over thousands of years by studying the rocks the glaciers leave behind. The maps below show the Arctic ice sheet as it stands today over North America and the position of ice at the height of the last ice age 18,000 years ago.

Present day

Last ice age

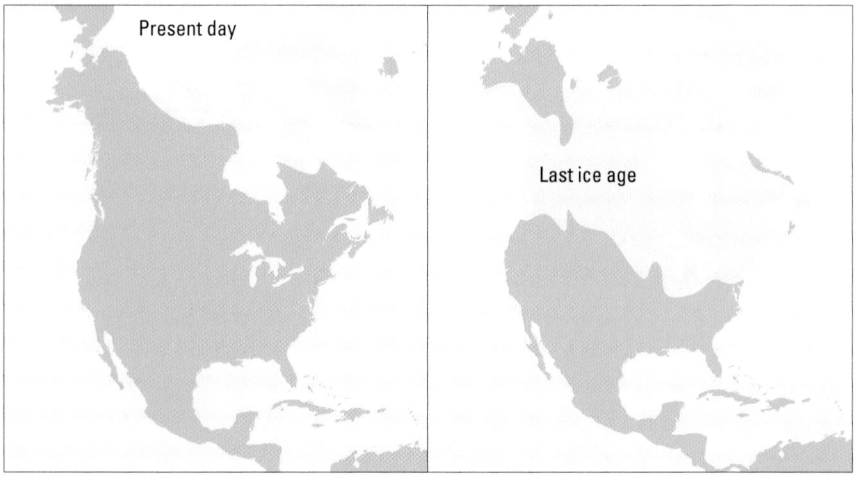

Chemical methods

Oxygen atoms exist in two forms, called isotopes, that differ in the number of their neutrons and so have different atomic weights (see ISOTOPES). Oxygen-16 (O-16) has 8 protons and 8 neutrons, while oxygen-18 (O-18) has 8 protons and 10 neutrons. Climatologists have discovered that the ratio of O-18/O-16 in glacial ice is higher when the ice forms under warm-air conditions than cool. The oxygen isotope ratio in successive ice layers in a glacier or ice sheet is an indicator of air temperature over time. Researchers have obtained ice cores to depths approaching 2 miles (3.2 km) in Antarctica and in Greenland, providing a climate record that extends more than 160,000 years. Paleoclimatologists date ice cores using a variety of methods, including linking the dust content at specific depths with known geological events and by correlating ice-core cold and warm peaks with other paleoclimate data.

Oxygen isotope ratios in the chalky skeletons of plankton show a reverse trend compared with those in ice cores: high O-18/O-16 ratios are correlated with cool-water conditions. Analyzing the oxygen isotope ratios in fossil plankton skeletons in marine sedimentary strata provides a climate record extending back more than 65 million years—when dinosaurs became extinct.

In ice cores taken from Antarctica and Greenland, carbon dioxide and methane gas levels in ice-trapped air bubbles usually show a close correspondence with air temperature: higher levels of the two gases correlate closely with higher temperatures.

Tree rings

Cut through the trunk of a tree and one of the most obvious features is the presence of concentric growth rings. One tree ring represents a year's growth, with faster cell growth and division in warmer months and a near-cessation in winter months. In northwest and central Europe, for example, the width of tree rings in oak trees is closely correlated with annual variation in rainfall rather than temperature. In wetter years, growth is greater. Inferred climatic records from tree rings extend back over more than 150 years for living oak trees and over 10,000 years for oak tree fossils. Similar records exist for pine trees. Dendrochronologists (scientists who study tree rings) can age fossil trees using radiocarbon dating (see RADIOCARBON DATING) or by correlating tree ring events with other paleoclimatic data. For example, striking tree ring anomalies can be correlated with geological events such as major volcanic eruptions, which cause a depression of global air temperatures for a year or two. A volcanic eruption on the Greek island of Santorini in about 1600 B.C.E corresponds with frost rings found in bristlecone pines that were growing in the western United States at that time. A sulfur dioxide peak in Greenland ice cores corresponds with that time period. (Sulfur dioxide is a major gas emission of volcanic eruptions).

Historical and archeological records

Meteorologists have been keeping precise climate records for less than 150 years. Data from many locations worldwide enable them to detect small shifts in global climate by direct measurement. However, historical records and archeological evidence, such as carved friezes on ancient Egyptian tombs, offer important clues to climate change in historical times. For example, during the 17th century, northwestern Europe experienced an unusually cold spell, which has since been dubbed the Little Ice Age.

Long-term climate change

Paleoclimatologists have constructed a record of global climate, based on average temperatures and rainfall, extending from the present day to about 3.9 billion years ago, when the earliest known rocks were formed. During this time, there have been extended periods when the Earth was consistently warmer than the long-term average (hothouse periods) and times when it was consistently cooler (icehouse periods). Most paleoclimatologists recognize five icehouse periods, and Earth is experiencing a warmer time (interglacial) during the currently late Cenozoic icehouse period. Glacials, or ice ages, are cooler times during icehouse periods when glaciers and ice sheets advance and cover large portions of midlatitude continental regions. The last glacial ended about 15,000 years ago.

About 65 million years ago hothouse conditions existed. The average annual temperature at the equator was at least 3.6°F (2°C) higher than today, with the poles at least 36°F (20°C) warmer than nowadays. There were no polar ice caps, and dinosaurs—at least some of which were cold-blooded—could live in polar regions. Earth's climate shifted to icehouse conditions about 33 million years, and the last 1.6 million years has seen a further cooling overall. What has brought about these changes?

Assuming levels of solar radiation reaching the Earth's atmosphere have remained relatively constant, a variety of geological and biological events can explain long-term climate changes.

Landmass distribution

The shape, size, and distribution of landmasses has a marked effect on global climate. Continents drift as a result of the movement of massive plates that underlie Earth's oceans and continents (see PLATE TECTONICS). Over hundreds of millions of years, continents fragment and drift apart. At other times they collide and form supercontinents. Landmasses deflect heat-carrying ocean currents, and the center of large continents is a long distance from the influence of warming oceans and so tends to have colder winter temperatures. The balance between landmasses absorbing or reflecting incoming solar radiation depends upon where they lie on Earth's surface and their overall albedo (reflectiveness; see GAIA HYPOTHESIS). Icy continents at high latitudes are paler and more reflective than green or brown vegetation-rich continents at low latitudes.

Uplifting of land surfaces can help explain the onset of icehouse conditions. For example about 40 million years ago, the collision of the Indian subcontinent with Asia began to uplift the land to create a broad, high Tibetan plateau and the nearby Himalayas. These new high-altitude obstacles would have altered air circulation patterns.

Volcanic activity

On timescales from a few years to millions of years, the level of volcanic activity on Earth's surface influences climate. Volcanoes release carbon dioxide, sulfur dioxide, and water into the atmosphere, all of which are greenhouse gases (see GLOBAL WARMING). Such gases absorb infrared radiation emitted from Earth's surface rather than allowing it to pass through Earth's atmosphere and into space. Higher concentrations of greenhouse gases have a warming effect by trapping more heat energy in Earth's atmosphere. At the end of the Permian period, about 250 million years ago, higher levels of volcanic activity could explain the enhanced hot-house conditions triggered at that time.

Carbon-fixation and carbon sinks

Organisms that photosynthesize—trap sunlight to manufacture foods from carbon dioxide and water—extract carbon dioxide from air or water. At times in Earth's history photosynthetic organisms removed carbon dioxide that subsequently became trapped in forms below ground. The resulting periods of low atmospheric carbon dioxide levels had a slight cooling effect on global climate. One such correlation occurred during the Late Carboniferous period, 295 to 324 million years ago, when vast swamps covered the supercontinent Pangaea on which grew giant clubmosses and other treelike plants. When the plants died, they became buried in oxygen-poor

A COOL, YOUNG SUN

Astronomers and physicists calculate that 4.6 billion years ago, at about the time Earth formed, the Sun emitted less solar radiation than it does today. About 3.8 billion years ago, when Earth's first watery oceans formed, the average temperature on Earth's surface should have been more than 36°F (20°C) lower than it is today, in which case the oceans would have frozen solid. However, at that time photosynthetic organisms (those that absorb carbon dioxide and release oxygen) did not exist, and Earth's atmosphere would have contained more carbon dioxide than it does today, and an enhanced greenhouse effect would have trapped more infrared radiation in Earth's atmosphere, warming Earth's surface. This enhanced greenhouse effect counterbalanced the weaker levels of solar radiation, so that temperatures remained above freezing.

A CLOSER LOOK

conditions within the swamp and so did not decay. The decay process would have caused the carbon inside the remains to be released as carbon dioxide. Instead the remains became preserved as coal. Major glaciations occurred in the Southern Hemisphere during the same period and may be linked to a reduced greenhouse effect.

Short-term climate change

During the current icehouse period, which began about 1.6 million years ago, there have been at least five major episodes of glaciation with many more minor ones, all separated by interglacial periods. In the last 15,000 years alone, since the end of the last major glaciation, although overall there has been a warming trend, it is punctuated by periods of cold. For example, a cooler period called the Younger Dryas (named for an Arctic flower that became widespread at that time) peaked about 10,500 years ago. A smaller cooling episode—the Little Ice Age—extended between the 16th and 18th centuries, during which time glaciers in the Alps advanced, and canals in Holland froze most winters. Conversely, there have been at least two unusually warm episodes in the last 15,000 years. The first, called the Holocene Climatic Optimum, peaked between about 5,000 and 6,000 years ago, when global annual temperatures were typically 3.6°F (2°C) warmer than today. The increased evaporation led to greater precipitation, which probably fueled the agricultural expansion that took place in the area known as the Fertile Crescent and helped establish the first civilizations in Mesopotamia (now Turkey, Syria and Iraq). During the Medieval Warm Period that peaked between the 10th and 11th centuries, Vikings from Scandinavia were able to colonize southern coastal regions of Greenland and establish agricultural communities there. Today the climate is too cold for agriculture. In summary, even when the climate is trending toward warmer or colder conditions over thousands of years, there are considerable fluctuations in temperature and precipitation in the meantime, which have major implications for people and other forms of life on Earth.

The explanations for short-term fluctuations in global climate include those that explain longer-term fluctuations, with some additional factors. A short episode of intense volcanic activity can cause global cooling, as happened in 1815 when Mount Tambora in Indonesia erupted. Dust that spread around the world and blocked some of the Sun's rays and caused snowfall across northwest Europe in the spring and low temperatures throughout Europe's summer.

Evidence from fossil plankton in sediments beneath the North Atlantic Ocean suggests that ocean currents can change strength and direction quite abruptly, so creating an equally abrupt climatic change. The Gulf Stream that crosses the North Atlantic from west to east is an unusually large, warm ocean current that maintains comparatively mild temperatures in northwestern Europe (see GULF STREAM). The plankton fossil record of cold- and warm-water species reveals that the Gulf Stream probably changed direction in an episode lasting several hundred years between 10,000 and 11,000 years ago, and the average temperatures in Europe fell. Melting glaciers in and around Greenland would have created a layer of freshwater floating above saline water in the North Atlantic, so disrupting normal oceanic circulation patterns. Paradoxically, global warming triggered cooling in northwestern Europe, something scientists fear might happen again should the current trend of warming continue.

Sunspots are regions of strong magnetic field on the Sun's surface. The number and size of sunspots undergo periodic fluctuations that are correlated with changes in the levels of solar radiation reaching Earth and the consequent changes in global climate. Essentially, greater sunspot activity is usually associated with higher solar radiation levels and slightly raised surface temperatures on Earth. There appear to be several cycles of sunspot activity superimposed on one another, producing peaks at about 11-year and 80-to-100 year intervals. Cold winters during the Little Ice Age of the 16th to 18th centuries are correlated with years of reduced sunspot activity, as measured in preserved wood remains by a decline in the annual yield of carbon-14 (C-14) that is produced by cosmic ray activity (see COSMIC RADIATION).

T. DAY

See also: ALVAREZ, LUIS; CLIMATE; COAL AND COAL GAS; EXTINCTIONS; FOSSIL RECORD; GAIA HYPOTHESIS; GEOCHEMISTRY; GEOLOGY; GLACIERS AND GLACIOLOGY; GULF STREAM; ICE AGES; ISOTOPES; PALEONTOLOGY; PLATE TECTONICS; RADIOCARBON DATING.

Further Reading:

Alley, R. B. 2002. *The Two-Mile Time Machine: Ice Cores, Abrupt Climate Change, and Our Future.* Princeton, N.J.: Princeton University Press.
Saltzman, B. 2001. *Dynamical Paleoclimatology.* New York: Academic Press.

THE DEMISE OF THE DINOSAURS

Some scientists (see ALVAREZ, LUIS) point to a meteorite impact as being the main cause of a dramatic climate change that caused the extinction of the dinosaurs some 65 million years ago (see EXTINCTIONS). The impact of a 6-mile (10 km) wide meteorite would be sufficient to raise vast amounts of dust into the atmosphere. A combination of factors—the blocking of the Sun's rays causing cooling and plant deaths, followed by the release of greenhouse gases causing global warming and then acid rain—would be sufficient to cause a mass extinction event. Evidence for the meteorite's likely impact site has been found in the Gulf of Mexico, coupled with an iridium-rich band in sediment dating to about 65 million years ago (iridium is an element rare on Earth but common in meteorites). However, more recent evidence suggests that the meteorite impact does not exactly correspond with the date of many of the dinosaur extinctions. Other factors, such as enhanced volcanic activity causing global warming, were taking their toll before the devastation of the meteorite impact.

A CLOSER LOOK

INDEX

Bold words with bold page numbers indicate complete articles. Italic page numbers refer to illustrations or other graphics.